The Greenie's Guide to the End of the World

Theodore David McCall

ATF THEOLOGY

Adelaide

2011

The Greenie's Guide to the End of the World:

Ecology and Eschatology

Theodore David McCall

National Library of Australia Cataloguing-in-Publication entry
Author: McCall, Theodore David.

Title: The greenies guide to the end of the world : ecology and eschatology / Theodore David McCall.

ISBN: 9781921817120 (pbk.)

Notes: Includes index.

Subjects: End of the world--Environmental aspects.
Ecology--Religious aspects.
Eschatology.

Dewey Number: 236.9

Cover design by Astrid Sengkey
Original painting by Tim O'Shea

ATF THEOLOGY

An imprint of the ATF Ltd
PO Box 504
Hindmarsh, SA 5007
ABN 90 116 359 963
www.atfpress.com

Contents

Section Two—Apocalypticism

Section Three—Some models of the eschatological transformation of all things

Acknowledgements

This book began life as a PhD thesis, so I firstly record my thanks to my supervisors, Denis Edwards and Phillip Tolliday. Their patient attention to detail and constant encouragement helped me realise that it was indeed possible to finish. Sometimes, perhaps often, we underestimate the quality and dedication of our theologians in this country. To two fine academics, whose minds I admire and whose theology I enjoy, I say thank you.

The wardens and people of St John's Church, Albany, WA (my parish until recently) are to be thanked. Many of the parishioners have expressed interest in the topic and all have been supportive of the undertaking and the time involved. The Diocese of Bunbury and particularly the recently retired Bishop David (who is also my father) must be thanked as well, for allowing and indeed encouraging clergy to take study leave for projects such as this. My interest in theology really began around the dinner table, and so, in thanking Dad, I also thank my Mum, Marion. Her prayers have not gone unnoticed. I also thank my extended family, all of whom have encouraged me in this undertaking in their different ways. Special mention must go to my sister Elizabeth and her husband Julian, whose genuine fascination with theology provided an unexpected boost for me in the streets of the cities of Italy.

Special thanks must also go to my friend and colleague, Ian Mackay, whose guidance and interest have also helped keep me going. A friend who has run the same race is one who knows the value of those who cheer from the sidelines. I thank all those other friends and colleagues, too many to name, who have helped in so many different ways: from friends whose company I enjoyed in the coffee shops of Albany, to colleagues and mentors in both the Diocese of Bunbury and the city of Adelaide, whose valuing of higher education in theology was one catalyst to my beginning this work.

I must also mention the support and encouragement of Hilary Regan and ATF Press, in particular Hilary's persistence in seeing this project through. I also thank the St George's Association for Education (based in my home parish of St George's Goodwood, SA). The Association was extremely generous in helping with the costs of publishing. In thanking the Association, I also thank the Rector of St George's Goodwood, Scott Moncrieff, and faithful parishioner of St George's and close family friend, Bill McDonald. Likewise, I thank two of my colleagues here at Saints: Tim O'Shea, for allowing me to use his striking painting on the front cover and Alex Smith, for assistance with the final copy for review and for introducing me to Robert Crotty, whose insights have been invaluable. Our Headmaster, Simon Murray, has also been tremendously supportive.

I wish finally to thank four precious people who have been with me on this journey from the start: my wife and closest friend, Alison, whose steady support and encouragement have been the source of much endurance; my daughter, Miriam, whose smiling face and guileless laughter are God's gift not just to me, but to many people; and my sons Alasdair and William, whose interested comments and perceptive questions have reminded me of theology's relevance to all people, not just theologians.

Theo McCall
St Peter's College, Adelaide
Feast Day of the Birth of Mary, Mother of our Lord, 2010.

Glossary

Aeon *or* Aevum	A description of time—not the absolute eternity of God himself—it is the relative eternity of the new creation, which participates in the absolute eternity of God.
Anamnesis	Memory in the strongest possible sense of recalling past events and people so strongly that they become a part of present reality.
Anthropocentric	Human centred, that is, looking at things from a human perspective and with a human focus.
Apocalyptic	Literally 'unveiling' but normally refers to the end of the world, whether destructive or transformative.
Apophatic	The understanding of God is found in what we do not say: an acknowledgement of the limits of human language and knowledge of God.
Christology	Knowledge of Christ, but particularly the theology surrounding Christ.
Consummation	The transformation of everything into God's new creation, preserving yet transforming.
Cosmology	A view of the world, that is, the cosmos, whether ancient or current.

Creatio Continua	God continues to create, whether through direct intervention or through the processes of the natural world (for example, evolution).
Creatio ex nihilo	God created the world out of nothing. This doctrine was particularly important in the early church in countering the Gnostic ideas of creation from pre-existing matter.
Diachronically	At the same time, across time.
Dipolar	Having two natures. Used particularly in this book as referring to God having two natures with respect to time: 'eternal' and 'immanent'.
Docetism	A heresy which viewed Jesus as divine but not human, or a spirit but not of the flesh.
Economic Trinity	The presence and activity of the Trinity within creation.
Ekklesia	The body of the church, the community of faith.
Entropy	The running down of energy: available energy becomes increasingly random and disordered.
Epistemology	The theory of knowledge, including its nature and scope.
Eschatological moment	The moment when God will gather all of time and space, the whole created order, and transform it.
Eschatology	The 'last things', normally referring to heaven, hell, purgatory and judgement. However, also refers more generally to the eternal destiny of everything.

Eschaton	The final transformation of everything into the new creation.
Exegesis	Interpretation of a text, that is, determining its meaning.
Gaia	A name given to the idea of the entire universe being inter-connected to the extent that it can be thought of as a whole: builds on ancient ideas of 'Mother Nature'.
Gnosticism	Heresy which views the material world as imperfect and knowledge (Greek: *gnosis*) as the ultimate form of perfection. The world is understood as being created from (imperfect) pre-existent material.
Immanent	God's presence in creation, usually in contrast to his transcendence.
Immanent Trinity	The inner, divine life of the Trinity: the relationship of the Father, the Son and the Holy Spirit within the God-head.
Kenosis	Self-emptying, referring to God's self-restriction in allowing creation to come into being or Christ's self-restriction of his divine powers in becoming human.
Material Inscription	A process in which the history of the cosmos, that is, the history of everything, is thought of being inscribed in the eschaton.
Millenarianism	Similar to 'apocalypticism', namely the end of the world in a very literal, physical sense.
Neo-Platonism	A philosophical world-view in which the world is viewed as imperfect, with ultimate salvation

coming from the salvation of the soul, as opposed to the body.

Omnipresence God's presence everywhere.

Ontology The essence of something: the nature of its 'being'.

Orthopraxis Right action: practical action which is 'correct' or 'orthodox'.

Panentheistic God's presence in creation and yet transcendence of it. God is both transcendent and immanent.

Parousia The second coming of Christ at the end of time.

Perichoresis Mutual indwelling. Two persons or beings dwell within each other mutually. They also participate in the attributes of the other.

Pneumatology Theology of the Holy Spirit.

Praxis Practical action.

Primordial Moment The moment of creation, when God chose to restrict himself, and from the restriction of eternity emerges 'the time of creation'.

Second law of thermodynamics The conversion of energy to heat (to do some useful work) inevitably means a decrease in the quality of the energy.

Shekinah A Jewish concept of God, describing God's presence with the people. The idea of the Shekinah included the sense of the Shekinah journeying with the people, and suffering with them.

Soteriology The theology of 'being saved' or 'redemption'.

Sublation Literally means the turning of one thing (minor)into something larger. However, in Ulrich Körtner's work it is a translation of the German 'aufheben' which means both to preserve and change. He uses the term solely with respect to apocalypticism to mean 'the Christian understanding of existence neither merely negates the apocalyptic understanding of existence nor shares it wholly without contradiction'. In other words, it describes Christianity's acceptance of apopalyptic thought, but acceptance which is shaped by the cross and resurrection of Christ.

Teleological Having a forward movement towards a specific Goal (Greek: *telos*).

Theodicy The study of the question of the presence of evil in a world created by a God who is good.

Transcendence God's nature and being is above and beyond anything in creation.

Introduction

The world that we inhabit, and the vast cosmos in which we exist, is incredibly beautiful and intricate. The advances of modern physics, including the latest telescopes and space exploration vehicles, have simply increased the sense of wonder which the ancient Jews wrote about in the psalms. The more I read about the delicate balance of the first moments of the cosmos, the more I marvel at the process which brought about the existence of the earth and the creatures that inhabit it. One of my joys in life is cycling. I live in the city of Adelaide, South Australia. It is a perfect city for cycling. When I first began this project, once or twice a week I would cycle in the early morning. The last part of the return trip would involve riding eastwards into the rising sun. On those occasions, as I felt the warmth of the sun on my face and was aware of its sheer power and energy, I was convinced, as I still am, that the universe is an overwhelmingly beautiful place, which God would never abandon or destroy. It is with this complete faith in God's love for his[1] creation that I set out to write an account of how this creation, with all its beauty and fragility, as well as its flaws and scars, might be transformed into God's new creation. Given the ecological crisis that we face, how can we meaningfully talk about the consummation of all things, without removing the impetus for ecological action? In other words, is it possible to develop an ecological eschatology? This book will begin the steps of reconciling ecology with eschatology, building on the belief that both are necessary in any balanced theology.

1. Some notes on language: firstly, throughout the work I have attempted to avoid the use of gender-specific language for God as much as possible, except for reasons of style or clarity or when directly quoting other writers. Where I have made use of the masculine personal pronoun for God, it is simply for the sake of style in those particular sentences. Secondly, the British convention of spelling has been followed throughout, likewise except when directly quoting other writers who follow the American convention.

1. Intimately connected

One of the assumptions of this work is indeed that the world is not only precious and loved by God, but that our very existence as human beings is fundamentally connected with the emergence of the universe. This may seem self-evident, but the sad reality is that many Christians, including many Christian theologians, have come to believe in a kind of neo-Platonism, which sees our pre-existing souls returning to a kind of bodiless existence in 'heaven'. Heaven in this context is defined rather imprecisely but may be understood as not being a part of the creation itself, which is transitional and contingent, but rather as an eternal paradise. In recent years this neo-Platonic approach has also been influenced by the New Age movement, which in turn has appropriated and modified the Hindu concept of re-incarnation. This work rejects these approaches, simply because the connection with our bodies and indeed the wider cosmos is a fundamental part of who we are as creatures and human beings.

A brief examination of current scientific thought, building on my experience and wonder of the rising sun, will show the connection between the cosmos and us. We are intimately and inextricably connected with the cosmos, at the deepest and most fundamental level. 'The cosmos is a unity. To understand ourselves we must understand the stars. We are star-dust—he ashes from long-dead stars.'[2] What Martin Rees is pointing out here is that at the basic chemical level of our bodies we are connected with the rest of the cosmos. Indeed a galaxy represents a vast ecological system. Inside the stars pristine hydrogen is transmuted into what Rees terms the basic building blocks of life—carbon, iron and the rest. 'Some of this material returns to interstellar space, thereafter to be recycled into new generations of stars.'[3] It is astonishing to consider that carbon atoms, those in every cell of our blood, in the trees around us, the furniture in our houses, 'have a pedigree extending back far earlier than our Solar System's birth 4.5 billion years ago. The Solar System itself condensed from the intermingled debris of many earlier stars.'[4] Even as I write, dust particles, circling newly forming proto-stars (which are not yet hot enough to ignite their nuclear fuel) are sticking together to make rocky 'planetesimals' which merge to make planets.

2. Martin Rees, *Before the Beginning: Our Universe and Others* (London: Touchstone, 1998), 19.
3. Rees, *Before the Beginning*, 17.
4. Rees, *Before the Beginning*, 18–19.

A theological reflection on this basic scientific account of life suggests that not only are we intimately connected with the rest of the universe, but that at a basic level we owe our very existence to the stars we take such delight in observing. To suggest, therefore, that we could exist apart from the cosmos seems to deny the importance of the cosmos for our existence. We are beings in relationship, not simply with our fellow human beings, but indeed with the entire created realm. Even to suggest that we could be *bodily* resurrected, without an understanding of the resurrection of all of creation, seems to down-play the significance of our relationship with creation. The charge of Gnosticism could be levelled at those who do not give sufficient value to world history and indeed the bodily matter of the cosmos itself. Tibor Horvath, writing within the Catholic tradition, suggests that any eschatology must include the consummation of *all* things.[5] Bodily resurrection, without any development of the doctrine of the consummation of all things, implies that we could exist without the rest of creation. But this approach ignores the intimate connection we have with all of creation.

2. Finely tuned processes

There is a further reflection which comes from an appreciation of contemporary physics: the processes which allowed the earth itself and then human beings to be formed at all are incredibly delicate and fine-tuned. If the conditions had been subtly different, then life, certainly as we know it, would never have formed. A few examples will suffice to illustrate the point.

The first is a simple point about the conditions on a particular planet that must exist for life to be able to form. Planets where life can evolve must, of course, have sufficient gravity to prevent their atmosphere evaporating into space. They must also be the right distance from a long-lived and stable star, so that they're not too hot or cold.

The second example is more nuanced. For a carbon nucleus to form (carbon being the critical base for life-forms) beryllium must form. Its

5. Tibor Horvath, *Eternity and Eternal Life: Speculative Theology and Science in Discourse* (Ontario: Wilfrid Laurier University Press, 1993), 148. For various Catholic perspectives on the question of bodily resurrection and the consummation of all things see Joseph Ratzinger, *Eschatology: Death and Eternal Life* (Washington DC: The Catholic University of America Press, 1988) and Bernard Prusak, 'Bodily Resurrection in Catholic Perspectives', in *Theological Studies* 61 (March 2000): 64–105.

nucleus contains four protons and four neutrons. To then form carbon another helium atom must be captured later in a separate collision. But this can only happen if the beryllium and helium can stick together quickly and easily, because beryllium made in the above manner is unstable and has a very short life-span. There has to be a resonance in the carbon nucleus and without this particular resonance carbon could not have been made in the stars. Somewhat ironically though, if the oxygen nucleus, which forms by carbon capturing a fourth helium nucleus, had a resonance 'the carbon would be processed into oxygen, or even further up the period table, as quickly as it was made'.[6] These particular features of carbon and oxygen, which seem to be simply accidents of nuclear physics, are in fact crucial for the wide-spread existence of carbon in stars and planets, and therefore for the course of cosmic evolution.

Two final examples will show that the conditions in the universe are perfect for the evolution of life and indeed must be this way for us to exist. Firstly, of crucial importance is the finely tuned balance between the effects of gravity and the kinetic energy of the inflationary universe. If gravity had been too strong at any point in the evolution of the universe, expansion would have ceased and the universe would have begun to contract. The galaxies would not have evolved as they did and therefore life would not have emerged. The ratio between the actual density of atoms per cubic metre and the critical density, which would cause expansion to cease, is denoted by the Greek letter Omega. The remarkable fact is that Omega doesn't differ enormously from a ratio of 1 today. 'There is not a tremendous imbalance between the effects of gravity and the energy of expansion—and this has striking implications for the early universe.'[7] Either expansion would have rapidly increased or gravity would have stopped the expansion completely. But our universe was given a very 'finely tuned impetus, exactly enough to balance the decelerating tendency of gravity'[8] and this is how the universe ended up as it is. Yet this finely tuned balance was achieved incredibly quickly. Very early on the expansion would have been accelerated exponentially, to enable the embryonic universe to inflate, before establishing the balance between gravitational and kinetic energy when it was only 10^{-36} seconds old.

Finally, the formation of stars, galaxies and planets depends on ripples or fluctuations in the early universe. The height of the ripples is described

---

6. Rees, *Before the Beginning*, 27.
7. Rees, *Before the Beginning*, 154.
8. Rees, *Before the Beginning*, 178.

by the number Q. If Q were too small the universe would be too smooth and remain dark and featureless. On the other hand if Q were too big, 'the cosmic scene would be dominated by black-holes rather than galaxies, and stars (even if they managed to form) would be buffeted too frequently to retain stable planetary systems'.[9] But Q was perfect for the evolution of the universe we now live in and which gave rise to our very existence.

The point of reflecting on the intricate balance of the evolving universe is that our existence is the result of an unimaginably long, complex and finely tuned series of events. Viewed through the eyes of a contemporary physicist, the creation is the result of a fantastic, continuing series of events, which will arguably never be fully understood. Of course *God's* perspective on creation is one we can only begin to imagine. Accepting, though, that assuming to understand God's perspective is the height of arrogance, if not idolatry, it is nevertheless within the bounds of a theologian's work to assert that God does not view the creation lightly. God delights in all of creation, including the processes of creation, which are intricate and finely-tuned. God would simply not dismiss creation or forget about it. If the argument for God's existence from the *design* of the world has been made redundant by Darwin's theory of evolution, there is still validity to the suggestion that the *processes* of creation evoke a sense of awe, appreciation and even love. We appreciate God as the author of these processes. But it is not unreasonable to suggest that God also delights in what he has created.

3. The consummation of all things

If we maintain that God loves the world, which is a fundamental premise of Christianity, then we must also maintain that God loves the very processes by which God brought about the world's existence. Because of the complexity of creation, because of its inherent beauty, because God has been intimately bound up in the long process of creating the world, it is difficult to believe that God would then abandon it or destroy it. It is difficult to believe that God would bring down God's wrath upon it, just as it is difficult to believe that God would abandon it to a slow decline into entropy.

This last point is crucial from a current scientific perspective. There have been various attempts to show that the entire universe may in fact be

9. Rees, *Before the Beginning*, 246.

infinite and eternal. The Russian cosmologist Andrei Linde has put forward the suggestion that the universe could indeed be eternal, but that it continually generates inflating regions which evolve into separate universes. This chaotic inflation, referred to by Martin Rees as the 'multiverse' theory has implications for eschatology. The suggestion that the universe is everlasting would seem to negate any need or desire for the ultimate consummation of all things, that is, the cosmos, although the question of theodicy might challenge that perspective—an unjust world, even an eternal one, would need transforming. Having noted the theory of chaotic inflation, or multiverses, it is important to point out that many physicists favour a model of the observable universe which would see it ultimately succumb to the results of continual expansion, condense into enormous black holes and then slowly decline into a state of low grade radiation. This is popularly referred to as the 'big freeze' as opposed to the 'big crunch' in which gravity finally overcomes the kinetic energy of the universe and the entire universe ends in a 'singular cosmic melting pot'.[10] Whilst this may indeed be the projection made by many physicists, mainstream eschatologists should nevertheless assert that this is not the final word. Whatever the Christian promise of resurrection ultimately might mean, the whole created order will be a fundamental part of that event. It is hardly a Christian position to assert that the 'big crunch' will have the final word. It is equally un-Christian, I am asserting, to suggest that the ultimate future for intelligent life, as the background temperature falls, will be to 'keep cool, think progressively more slowly and hibernate for long intervals.'[11] Eschatology, at least in all of its traditional forms, asserts that death, including I suggest the death of the universe, is not the final state of all things. Phillip Tolliday has expressed this position well:

> I think we can give all sorts of meanings to 'the end of the world' but I think the thing that most distinguishes the theological narrative from the cosmological one is that in the case of the theological story 'end' as full stop, no more isn't really applicable. Sure, it's the end of a period, the end of world time, of world history, perhaps of mortality, of finitude. But then comes a whole discourse about the kingdom of God; the new heaven and the new earth, a new creation—and

10. John Polkinghorne, *Science and Christian Belief: Theological Reflections of a bottom-up thinker* (London: SPCK, 1994), 162.
11. Rees, *Before the Beginning*, 209.

> whatever meaning we may give to these, it seems that it must be something other than nothing.[12]

Not all recent eschatologies share this conviction. Not only have some writers asserted that the resurrection does not have much to do with the cosmos and matter is unable to be brought to a final completion,[13] but others have asserted that, should the universe finally succumb to entropy, this would seem to deny our faith. Theologians as respected and well-known as John Macquarrie and Ted Peters have asserted this:

> Should the final future as forecasted by the combination of big bang cosmology and the second law of thermodynamics come to pass, wherein the law of entropy has the last laugh and the cosmos drifts into a state of irrecoverable equilibrium—and if some consciousness could then say, 'that's all there is'—then we would have proof that our faith has been in vain.[14]

John Macquarrie has expressed something very similar in *Principles of Christian Theology*:

> Let me say frankly, however, that if it were shown that the universe is indeed headed for an all-enveloping death, then this might seem to constitute a state of affairs so wasteful and negative that it might be held to falsify Christian faith and abolish Christian hope.[15]

It is significant, however, that both writers want to avoid separating redemption and transformation from creation itself. That is to say, both share my conviction that creation itself, by which I mean the entire created order—the cosmos—will be transformed in the eschaton. Perhaps

---

12. Phillip Tolliday, *On the End of the World* (Adelaide: Adelaide College of Divinity & Flinders University Unpublished Lecture, 2003), 6.
13. For a discussion of this issue see Horvath, *Eternity and Eternal Life: Speculative Theology and Science in Discourse*, 148.
14. Ted Peters, *God as Trinity: Relationality and Temporality in Divine Life* (Louisville, Kentucky: Westminster/John Knox, 1993), 175–176.
15. John Macquarrie, *Principles of Christian Theology: Revised Edition* (London: SCM, 1966 & 1977), 356.

the distinction between my position and Peters' in particular is that Peters wants to affirm that the transformation of the world will be apprehensible temporally.[16] For him eschatology is the link between the eternal and the temporal and the 'arrival of consummate eternity creates the wholeness of time'.[17] Certainly I would agree with the statement that all of time and all of creation will be gathered and embraced by eternity in the new creation. However, I reject the assertion that this must be temporally apprehensible and that the descent of the cosmos into a state of entropy would seem to deny the existence of God and the reality of Christian hope.

It is therefore from a conviction that God means to transform the world, for the reasons given above, that I want to begin developing an eschatology which takes seriously the fragile world and cosmos we call home. As I began writing this book, my nickname for the work (eventually to become the title proper) was *The Greenie's Guide to the End of the World*. In many ways this is an accurate title. That the world faces an environmental crisis is indisputable. That the world and its people need the hope of the resurrection is an assertion I would make. The task is to develop an eschatology which takes seriously the environmental crisis facing us and yet doesn't abandon the traditional hope of a transformed world.

4. Method

The book will take the following shape. I shall examine the work of several key theologians, who help shape an ecologically friendly eschatology. The book will build towards a model of ecological eschatology, using the work of these authors. I will undertake a critical theological assessment of each author's views. Thus my method of developing such a model will be cumulative, with the detailed analysis of each author adding to the construction of the model. Whilst no one author provides us with a complete model, each provides hints and suggestions, which help develop an ecological eschatology. It's important to note the inherent difficulty of constructing a fully developed eschatology, given the inevitable limitations of human knowledge as it relates to eschatology. This book has the more modest aim of setting out to put in place the crucial elements of a model of ecological eschatology.

16. Peters, *God as Trinity*, 177.
17. Peters, *God as Trinity*, 177.

A model of ecological eschatology must explain how creation might be eschatologically transformed, and how we might participate in that transformation. Such a model must therefore explain how God might transform creation: I call this the objective side of the model. The model should also explain how human beings and the rest of creation might participate in that transformation: I call this the subjective side of the model.

A model of ecological eschatology is desirable, because it could help people behave more responsibly towards the environment. Thus my hope is that a sense of orthopraxis may emerge from the model, in the sense that it evokes within the reader a desire to act in an ecologically responsible manner. So a model of ecological eschatology ought to encourage us not simply to hope, but also to perform acts of hope. We should be inspired to strive for righteousness, justice and peace now, even as we hope that these things will be established eternally.

The question of God's transcendence is crucial[18] in the development of the model; the objective side of the model is important, because, as I will show, the transformation of creation is *God's* action. However, the subjective side of the model is equally important. Any model of ecological eschatology must affirm the importance of the processes of the world, both human and non-human, in the final consummation of all things. Otherwise the question of the ultimate relevance of those processes emerges. I'm suggesting that, if we accept that the processes of the world, including for example the processes of evolution, are critical to our existence as we know it, then these same processes should be seen as having a participatory role to play in the transformation of creation.

Finally, my aim is to develop a model of ecological eschatology which will be able to be used, in the sense of being a theological basis for education and practical action, in parishes, dioceses and indeed the world-wide communions of churches. Consequently, for the purposes of this book, I will be restricting my focus to Christian theology. For academic theology to be usable in a parish or diocesan setting, it must at least be shown to have evolved from the Christian tradition, rather than be something

18. For a discussion of this issue see Mark Worthing, *God, Creation, and Contemporary Physics* (Minneapolis: Fortress Press, 1996), especially chapter five, 'Can God Survive the Consummation of the Universe?', 160–198. Worthing writes, 'Only the active presence of the transcendent God of Christian faith redeems the future of the physical universe. Attempts to posit an eternal universe without taking into account the almighty Creator have not been successful.' Worthing, 198.

completely and radically new. It must show clear and demonstrable links with this tradition.

5. The concept of memory

One of the central concepts in my work is 'memory'. The importance of memory is highlighted at various points in the work. However, I need to alert the reader to the fact that the concept of memory is used in various ways throughout the book. Sometimes it is understood to mean *our* remembrance of people. For example, in Elizabeth Johnson's work the concept of memory is specifically developed, with a focus on our remembrance of those people, saints as she terms them, whose lives would otherwise be forgotten. She discusses several different understandings of memory in her work. Whilst her focus is on human beings remembering other human beings, I will demonstrate that her work is easily adapted to include solidarity with all of creation and the remembrance of those parts of creation which have been victimised or destroyed completely.

In other places throughout the book the concept of memory is used in the sense of God's memory of us. Creation, including all human beings, will never be forgotten by God. God will remember creation, in the fullest sense of remembrance, meaning that creation will be present and alive in God.

In my work on Alexander Schmemann's theology the concept of memory is explicitly connected to the liturgy and the remembrance of Christ, particularly in the Eucharist. Human beings also remember all of creation, both in the celebration of the Eucharist and in other acts of remembrance, and in the process of remembering bring creation and themselves to God. Thus the act of remembrance, on the part of human beings, leads to human beings and the rest of creation entering into *God's* memory, which must be understood not as timeless eternity, but as everlasting life.

6. Structure

In section one (chapters one and two) I shall briefly analyse some ecofeminist critiques of traditional theology and especially eschatology. Catherine Keller has written at some length on what she views as the critical problems with traditional eschatological models. Keller considers that eschatology, particularly in its apocalyptic form, has not only not paid sufficient attention to the value of creation itself, but has helped promote a culture which is anti-environmental. Beginning with the book of Revelation, she has worked extensively on the theme of the sea, as a symbol of biodiversity. It is precisely the sea, which the book of Revelation describes as eventually dying and indeed being no more. Keller suggests that this is symptomatic of apocalyptic eschatology: there is no room for diversity, rather at the end everything will be gathered up into a kind of pristine uniformity. What I will suggest, however, is that Keller's critique focuses too closely on apocalyptic eschatology, reflecting perhaps in part her location as an American theologian—she develops at some length the link between Christian fundamentalism and industrial expansion. But more than simply focusing only on one branch of eschatology, Keller also fails to develop an alternative which is hopeful in the sense of ultimate hope.

No examination of feminist critiques of eschatology would be complete without an exploration of the work of Rosemary Radford Ruether. She is perhaps the best known eco-feminist to have addressed the topic. Ruether has been particularly critical of what she calls the male desire to dominate the world and to flee from it. It is out of a fear of the earth, the female and all that is not masculine that the desire to escape to a heavenly realm has arisen. She develops an alternative vision of what it means to be Christian in a world which is facing an ecological crisis. I shall critically evaluate her vision and show that it ultimately leaves us in despair.

Section two (chapters three and four) will be spent examining the contributions made by Ulrich Körtner and Jürgen Moltmann to the field of apocalyptic eschatology. Ulrich Körtner's work is perhaps relatively unknown in English speaking theological circles, but addresses many of the critical issues in this field. Importantly Körtner categorises apocalyptic thought in its various subsections. This reveals that Keller and Ruether have not necessarily understood some of the subtle differences between various kinds of apocalyptic traditions. Jürgen Moltmann is arguably the twentieth century's greatest contributor to the field of eschatology. His work on apocalypticism and apocalyptic eschatology (he makes a distinction between those two terms) is part of his overall contribution to eschatology. The question which is being asked by examining these two theologians is: 'Can apocalypticism contribute to the vision of a transformed cosmos, whilst not denying the importance of ecology?' This is a pertinent issue for the development of an ecologically friendly eschatology, because of the criticisms levelled at apocalyptic world scenarios by writers such as Keller and Ruether. An analysis of Körtner and Moltmann reveals that not all forms of the apocalyptic tradition are as anti-ecological as Keller and Ruether suggest.

Nevertheless, what I hope to show is that, taken in isolation, apocalypticism, in its various forms, is not able to provide us with a model for ecological eschatology. Other contemporary eschatologies help provide us with a more embracing model, which leads me to critique the work of John Polkinghorne and Ernst Conradie in section three (chapters five and six). Polkinghorne's model of two worlds, the old creation and the new creation, existing parallel to each other, but linked by projections from the old to the new,[19] allows us to consider exactly how the old and new creation might be related to each other. Conradie also develops his eschatology along similar lines and suggests that a concept of 'material inscription' may help us understand the relationship of this world to the world to come. Events throughout history are inscribed into the new creation and indeed form a critical part of that creation. Polkinghorne and Conradie's similar theological concepts help shape the objective side of the model of ecological eschatology, that is, how God might transform creation. However, in the case of both writers, I will argue that their models do not allow for significant *active participation* by human beings and the rest of creation in God's eschatological transformation of all things. That is to say, Polkinghorne and Conradie help in the development of the objective side of the model, but not the subjective side.

The book concludes in section four (chapters seven and eight) with an examination of two writers whose work gives us the final pieces of the puzzle. The work of Elizabeth Johnson and Alexander Schmemann is able to be used as a way of understanding how all of creation might participate in the final consummation of all things: the subjective side of the model. I will discuss how their approach to eschatology might indeed provide the model for understanding the active participation of all of creation in God's transformation of the cosmos *into* the eschaton. This is a development of Conradie's idea of the 'material inscription' of the cosmos, but from a more participatory perspective. Johnson is another feminist writer, but one who considers that the traditional eschatological models are not beyond redemption, even given the current ecological crisis and her own feminist convictions. As mentioned above, a part of her theology is the use of memory. She provides four examples of women's memory: 'recovering lost memory', 'rectifying distortion', 'reassessing value' and 're-

19. See John Polkinghorne, *The God of Hope and the End of the World* (New Haven & London: Yale University Press, 2002), 121.

speaking the silence'. She also develops the template 'Narrative memory in solidarity' with three categories: 'subversive memory', 'critical narrative' and 'solidarity in difference'. Her work on memory, and its power, helps us consider how even tragic events in history, and, I will argue, in creation itself, may be redeemed and brought before God in the memories of the living. This model, which, as I will show, finds particular expression in the Eucharist, allows us to cope with those parts of creation, including what Moltmann rather provocatively calls the 'victims' of evolution, which have been exploited, damaged or destroyed. Nothing will be lost, but all will be redeemed and, crucially for the development of an ecological eschatology, we have our part to play. Thus the theme of memory or remembrance gives us a hermeneutical key to ecological eschatology. The processes of creation and indeed all of creation itself will never be forgotten, but will be remembered by God in the fullest, most complete Jewish understanding of remembrance. Because God remembers all of creation, in all of its diversity and with all of its faults, what we do now for its preservation ultimately does matter. What we do now is of eternal significance.

Our participation, and that of all of creation, finds expression in liturgical theology. Schmemann's work on liturgy and symbolism, whilst not that of an ecological theologian *per se*, can be used nevertheless as a way of understanding how we might participate in the transformation of all things. Through the offering up of all of creation to God, and giving thanks for it, particularly in the Eucharist, we help make manifest the kingdom of God, through the Holy Spirit. This is possible, because the world is understood as a sacrament. In our liturgy, when we bless, or thank, God for his creation, we are giving back to God what God has created. This enables the world to regain its true identity, as belonging to God and existing to sing God's praises. This liturgical act of giving to God what belongs to God, of returning the matter of life to God, transforms that matter into the material of the kingdom of God. To use Conradie's term, God inscribes it materially, but, unlike in Conradie's model, our participation is crucial. To extend Schmemann's thoughts further, this action is not something restricted to human beings or to the celebration of the Eucharist. The Eucharist is a symbol, and the climax, of what is happening in a cosmic liturgy. The things of God are being returned to God, and are thus transformed into the things of the kingdom of God. We play our part, as does all of God's creation, in the cosmic song to God's glory.

Section I

The challenge of feminism

Chapter One
Catherine Keller

1.1 Introduction

That the world is on, or at the very least approaching, the brink of environmental disaster seems a fairly obvious point to anyone who cares about the natural world. From a theological point of view, partly in response to Lynn White's now famous article[1] but mostly in response to the increasing pressure of environmental issues, much good work has been written with ecology in mind. Unfortunately some, though not all, of this theology has worked from the basis that traditional models of theology, particularly of eschatology, are no longer relevant in our current environment of ecological degradation. A number of eco-feminist writers in particular have asserted that traditional theological models are indeed partly to blame for the current ecological crisis. Some, though by no means all, have sought to develop entirely new models of theology, often discarding the bulk of what has gone before them, in an effort to make theology relevant to the current crisis. Yet the sad reality is that these particular models have thrown out, often quite deliberately, some of the hope of the gospel. This section will examine some of these alternative models, firstly as developed by Catherine Keller, with the main focus of chapter two being on the work of Rosemary Radford Ruether. I will critique their work, particularly from the point of view of Christian hope, but also from the perspective of ecological ethics.

At this stage I would also like to point out that my work is not intended to be critical of feminist theology in general. My penultimate chapter looks at the significant contribution of Elizabeth Johnson, another feminist theologian, to the field of eschatology. This section is rather meant as

1. Lynn White, Jr, 'The Historical Roots of Our Ecologic Crisis,' in *Science* 155 (March 1967): 1203–1207.

a critique of the *eschatology* of two feminist theologians, Keller and Ruether. In examining their work, and later that of Johnson, I will attempt to show that the valid feminist criticisms of eschatology in general, and apocalypticism in particular, need not necessarily void the essential hope of traditional Christian eschatology.

1.2 Criticising apocalypticism

One of the good contributions from eco-feminist theologians in particular has been to criticise the apocalyptic tendencies within the Judeo-Christian tradition and the destructive attitude to the environment which they engender.[2] There has been a process of conditioning within Christian civilisations, suggests Catherine Keller, which has resulted in the participants in those civilisations viewing the world as ultimately expendable:

> Participants in Christian civilisation, which extends far beyond the bounds of belief, have been programmed by ancient visions to expect that when the going gets rough, the world will go.[3]

The destructive tendencies of our modern Western civilisations, suggests Keller, have their root in the belief that ultimately the end of history will come. This apocalyptic assumption, she suggests, under-girds everything we do at a national and international level. Because of the sense of the futility of resistance in the face of ultimate apocalyptic destruction and the lack of belief in truly sustainable options, our modern state has become dependent upon ecological and military exploitation to sustain our middle-class life styles. Needless to say, these life-styles carry an enormous cost to the environment and indeed developing nations. Indeed we have been led to the threshold of annihilation. But the impetus to change our lifestyle, and the destructive methods we use to maintain it, has been blunted by what Keller calls the myth of the Apocalypse. The expectation of an end-time and the end of time has defined the limits of our consciousness. Despair and the sense of ultimate futility rob us of the impetus

2. See for example, Anne Primavesi, *From Apocalypse to Genesis: Ecology, Feminism and-Christianity* (Minneapolis: Fortress Press, 1991).
3. Catherine Keller, 'Women against Wasting the World: Notes on Eschatology and Ecology', in *Reweaving the World: The emergence of Eco-feminism,* edited by I Diamond and G Orenstein (San Francisco: Sierra Club Books, 1990), 251.

to put a stop to the destruction. Yet this despair is complemented by the fervent hope for the end; the myth of the Apocalypse contains the fundamental hope of a new beginning. Whatever destruction is wreaked upon the earth as a result of the ecological destruction caused by Western consumerism will be forgotten, along with the earth, because God will make all things new.

Nuclear destruction remains another frightening possibility, which is favoured by some, claims Keller, because of its capacity to fulfil the prophecies of Revelation (for example Revelation 16:17f). But the possibility of nuclear destruction is simply the sister of the emerging reality of ecological melt-down. Both nuclear annihilation and ecological destruction have at their source the connection, which Keller identifies, between the myth of the Apocalypse and the flagrant waste of our world's resources due to Western consumerism and our utter contempt for the environment.

1.3 Signs of hope

Keller quite rightly sees signs of hope in our contemporary society. Both as individuals and as a society as a whole we are becoming more aware of the impending ecological disasters. The awareness of the global warming effect of the burning of fossil fuels, and the accumulation of industrial waste gases, has increased substantially in the last few decades. The awareness of the global environmental effects of our life-style has dramatically affected the way we live our lives. Keller makes the interesting observation that simple conversations about the weather reflect the daily destruction of the physical environment. Local people in the state of New Jersey (Keller's state of residence at the time of writing 'Women against Wasting the World') comment on the unusually warm June weather. The evidence that this is due to the effects of global warming is mounting.

Keller is of course correct in asserting that our daily lives are directly affected by the ongoing destruction. For example, at an individual level, every summer in Australia we are reminded of the dangers of being exposed to the sun for long periods. Those who are going to be spending more than 10 or 20 minutes in the sun without a hat or sun-screen take a substantial risk, which simply didn't exist to anywhere near the same extent in the first half of the twentieth century and before.[4]

4. One interesting response at a corporate and international level to global warming has been the search for more environmentally friendly motor cars. At the time of writing two major manufacturers (Honda and Toyota) are offering 'hybrid' cars for sale in

Importantly though, whilst some of the current ecological damage may, unfortunately, be irreversible, the avoidance of a total ecological catastrophe will only occur if environmental action is sustained. Keller asserts that sustainable wide-spread action depends on more than simple dispersion of information and public acceptance of that information. At a basic level each of us could live with the prospect of a permanently scarred planet, if the destruction of the entire eco-system could be avoided. Yet, she suggests, within each one of us there is something which seems to give up on the future, each time yet another piece of ecologically bad news emerges. It may therefore be time to name the 'apocalyptic element that has embedded world destruction within a vision of Divine providence moving history from creation to conclusion. The mythic miasma of a few apocalyptic texts, operating out of context and unconsciously, seems to preform our sense of time and history.'[5] In naming it, the connection between the myth of the Apocalypse and the destructive tendencies and sense of despair and futility which pervades Western capitalism might be broken, which in turn would lead to and allow sustainable environmental action.

1.4 Naming apocalypticism as a problem

What Keller therefore seeks to do is indeed name the destructive, despairing fundamentalist approach to the Apocalypse which suggests that there is no other way than regeneration through destruction. It is this connection between ecology and eschatology[6] which she explores in an effort, not to ridicule or angrily discount the dominant, militantly evangelical voices which believe that salvation arises out of destruction, but rather to hear them and then develop an alternative.

Early Christians lived with high expectations of Christ's imminent return and of a new heaven and a new earth, in which the old order would be transformed, wars and exploitation of all kinds would cease and the world would live in harmony. The heavenly Jerusalem was the symbolic representation of this new world order, the in-breaking of the Divine.

Australia as financially viable alternatives to purely petrol cars. This visible response and major financial investment on the part of these two giant Japanese manufacturers indicates a growing awareness in the market-place to the crisis.

5. Keller, 'Women against Wasting the World', 253.
6. See also Catherine Keller, 'Eschatology', in *Dictionary of Feminist Theologies*, edited by L Russell and J Clarkson (Louisville: Westminster John Knox Press, 1996), 86.

Apocalyptic eschatology was a radical development of this hope, with its expectation of a catastrophic end of history and the triumphant return of the Messiah and a final judgement, in which the saved join Christ in triumphant glory and the 'sons of darkness' (as Keller calls them here[7]) are condemned to eternal corporeal punishment. Whilst mainline Christianity quickly moved away from these more radical developments of apocalypticism and towards a tamer and more individualised expression of eschatology, Keller argues that their return in our time suggests the return of the repressed. She disputes that apocalyptic eschatology was a marginal exception.

Whether or not Keller's theory about apocalyptic eschatology being more than a marginal exception is correct, there is little denying her assertion that the industrial world, particularly as represented by the might of US industrial power, has taken a dualistic and hierarchical approach to nature and its resources. She is convincing in her argument that both the secular exploitation of nature and the apocalyptic visions of the future drive towards the end of time and the world, because neither 'respects the spatiotemporal rhythms of earthly ecology'.[8] There is a separation made between human beings (or, more particularly from Keller's feminist perspective, 'man') and the natural world, just as, theologically, there is an asserted distinction between an absolutely transcendent Deity, ruling from outside his creation and utterly independent of it, and that creation. This separation yields a temporally dualistic understanding of a beginning (creation) and a literal end (the apocalypse).

Within this context, as we shall see in the writing of both Ruether and Elizabeth Johnson (as different as the writings of those two feminist theologians are in many other respects), the female becomes associated with bodiliness and with nature, which is subject to externalisation and subjugation. 'Woman, as whore, old wife, witch, is the embodiment of time, which is to be used up, which is running away'.[9]

The dualistic framework within which this theology operates has found a convenient bed-fellow in the mechanised scientific understanding of the world which has, until very recently, dominated our thinking for several centuries. In this paradigm nature is something to be exploited, because we are separate from it and can observe it and use it independently of our own involvement or participation within its processes. We

7. Keller, 'Women against Wasting the World', 255.
8. Keller, 'Women against Wasting the World', 255.
9. Keller, 'Women against Wasting the World', 256.

have a responsibility to probe nature's secrets and to exploit them. Nature is understood in this philosophy dating from the Enlightenment period as a 'thing', a machine which can be taken apart, examined, and put to work in new ways. Its only value lies in its ability to enable the progress of human beings to continue.

Keller's key point, based on her critical reading of the book of Revelation in particular, but also on the philosophical approach to nature which I outlined in the previous paragraph, is that there has been an overwhelming desire to simplify things. Pristine simplicity is the goal of the transformed creation, symbolised powerfully and strikingly in the image of the New Jerusalem being 'pure gold, as transparent as glass' (Rev 21:21). So the debasement of nature, and all things associated with nature (including women themselves, suggests Keller) is 'parallel to the glorification of all that eludes it, God and man'.[10]

1.5 The sea

The particular symbol, which Keller interprets as representing nature, is the sea. The sea represents a highly evolved and delicately complex relationship of organisms. It is representative of nature exactly because of its complexity. And it is this complexity and diversity which is being eluded in some eschatology when it attempts to elude nature and indeed exclude it from the eternal bliss that the 'sons' of light enjoy. Keller also correctly points out that the Hebrew scriptures contain several references to the sea as being a monster needing to be contained. The 'deep' was feared. The deep was what God overcame and overthrew in the creation of the world.

There are a number of references to this idea of God overcoming the deep in the creation of the world, which are worth examining briefly. Firstly, at the start of the first account of creation in the book of Genesis (Gen 1:1–2:4a), the earth is a formless void and darkness covers the face of the deep, whilst a wind from God hovers over the face of the waters. It is this deep, this formless chaos of the world, which God ultimately shapes into order, by commanding that there be light, a dome to separate the waters, dry land to separate those waters now beneath the dome and so on.

The book of Job and some individual psalms also reflect this myth of conquering the sea. During his response to Job out of the whirlwind God questions Job, as to whether he was the one who put a limit on the sea:

10. Keller, 'Women against Wasting the World', 256.

'Thus shall you come, and no farther, and here shall your proud waves be stopped' (Job 38:11). The psalms too reflect this understanding of the sea. The deep, or the waters, are restricted by God. A limit is placed by God which the waters cannot pass: 'You set a boundary that they may not pass, so that they might not again cover the earth' (Ps 104:9).[11] It is also from the watery depths that God rescues God's servants (Ps 18:4 & 16).

What is interesting, however, is that in the first account of creation, God actually creates the great sea monsters and every living creature which swarms in the sea. God is the creator of all things, which is of course exactly the point the Priestly author(s) was trying to make in compiling this account of creation. In response to the crisis of exile in Babylon, the author of this first account of creation writes down a hymn of praise (probably pre-existing in some form) which proclaims that the God of Israel is in fact the Creator of the world.

Keller suggests that this concern with defeating and ultimately eliminating the sea arises out of the Jewish inheritance of the 'old Babylonian identification of the sea with the primordial; with the female, chaos, the Tehom'.[12] Because of this association it is the sea which will pass away, in the vision of the new heaven and the new earth given in the book of Revelation:

> Then I saw a new heaven and a new earth; for the first heaven and the first earth had passed away, and the sea was no more. (Rev 21:1)

Keller argues that this 'tehomophobia', as she terms it, suggests a systematic rejection of fluidity, indeterminacy and uncertainty. There is a rejection of nature's transience, of its chaos and therefore also of its creative complexity:

> Like an old, used-up wife the consumed matrix of life is cast aside. Into the space, the *outopos* which is 'no place', descends the virginal new order. An urban architecture supersedes the organic topologies of earth and sea. The text thereby pulls

11. Biblical quotations and references throughout this book are taken from the *New Revised Standard Version Bible* (National Council of Churches of Christ in the United States of America, 1989).
12. Keller, 'Women against Wasting the World', 256.

> the plug on sea dragons, draining from the universe once and for all the salty fluids of our bioplasmic genesis.[13]

Otherworldly eschatology has changed the earthy, historical Hebrew faith into 'a fantasy of final escape from transience and death'.[14] Here she is highly critical of Moltmann, suggesting that hope (Moltmann's key focus in eschatology) remains as much a part of the problem as the solution and that Moltmann wants to have his nature and escape from its transience too.

Keller's work here is a fairly brief and arguably simplistic analysis of Moltmann's theology. Later chapters of this book will examine Moltmann's eschatology in greater detail, but suffice to say, whilst criticism can be made of his work in the area of evolution, there is no doubting his commitment to the salvation *of* nature (as opposed to salvation *from* nature). Keller fails to appreciate the nuances of his work, in which it is *this* world which is transformed. Certainly we are to be redeemed from decay and death, but creation is an integral part of the promised eschaton. My work on remembrance as a hermeneutical key to ecological eschatology will suggest, as a development of Moltmann and Elizabeth Johnson's theology, that creation with all its complexity, will be gathered up into God.

It is precisely this complexity, though, which Keller sees as being rejected by eschatology as it has developed over the last 2000 years of theology. There is a drive towards unity and away from chaos, which is personified by the traditions of the evil sea monster. Transcendent unity, which is a 'profound impetus in all patriarchal spirituality',[15] is achieved at the expense of nature and multiplicity.

There is little doubt that, regardless of whether Keller's theory about the relationship between Western industrialist capitalism and apocalyptic theology is correct, the natural world is rapidly losing much of its diversity due to the consequences of ecological degradation. Our soil is eroding, many species of plants and animals are disappearing, the forests are being destroyed at an alarming rate; the diversity of life on earth is being reduced.

---

13. Catherine Keller, *Apocalypse Now and Then: A Feminist Guide to the End of the World* (Boston, Massachusetts: Beacon Press, 1996), 80.
14. Catherine Keller, 'No More Sea: The Lost Chaos of the Eschaton', in *Christianity and Ecology: Seeking the Well Being of Earth and Humans*, edited by D Hessel and R Radford Ruether (Cambridge, Massachusetts: Harvard University Press, 2000), 185.
15. Keller, 'Women against Wasting the World', 257.

1.6 Warnings from the book of Revelation

One interesting and tragic occurrence in all of the ecological destruction we see occurring around us is that the sea itself is suffering the effects of long term abuse. Given the anti-sea perspective which Keller reads in the book of Revelation, this gradual destruction of the sea and its diversity is not without irony. Without wanting to suggest that the book of Revelation is becoming literally true, it nevertheless is possible to 'read the contemporary plight of the earth's real oceans as a literalisation of the apocalyptic tehomicide'.[16]

Keller develops a fascinating line of argument here. Although no-one would accuse her of being a fundamentalist, she points out in a convincing manner that the kind of destruction that the book of Revelation envisages occurring to the sea is in fact happening. The sea is described as becoming blood with a third of the creatures of the sea dying (Rev 8:9). Having read this, Keller then invites us to consider the increasing incidence of 'red tides', that is toxic algal blooms, which 'release deadly toxicants into the water, killing other marine life and poisoning people'.[17] Although Keller does not wish to go premillennial, as she terms it, nevertheless there is a sense in which she makes a significant connection between the apocalyptic vision of St John the Divine[18] and the current ecological crisis, particularly as seen in the conditions of our oceans and seas. Eventually, according to the vision of John, every single living thing in the sea dies (Rev 16:3).

Keller's work, as she has developed this line of thinking with respect to the sea, has been concentrated on the book of Revelation. However, she does briefly examine the broader biblical picture of the sea. The book of Job serves as her example of this alternative biblical portrayal of the sea. Whilst it is true that God's response to Job includes this sense of setting boundaries for the sea (Job 38:11), there is no sense of the sea itself being in any way a negative part of creation. It may well represent chaos, which needed to be contained by God, but, in the context of chapters 38–41 of the book of Job, this passage is as much a celebration of chaotic creation as it is a celebration of God's creative activity. Even the awesome description of the sea monster Leviathan, in which God describes Leviathan's mighty strength and its impenetrable armour, is a celebration of creation, which

16. Keller, 'No More Sea', 185.
17. Keller, 'No More Sea', 185.
18. Whilst acknowledging the complexity surrounding the question of the authorship of the book of Revelation, for simplicity's sake I refer to the author as John.

cannot be contained by human beings nor controlled by them. Yet there is nothing negative or sinister in God's description of this fierce creature. Indeed, as Keller puts it:

> The text offers a rare witness to God's bottomless enthusiasm for the wild and the wet—in particular for the remarkable achievement of the complex organisation of this formidable organism.[19]

The broader biblical picture of Leviathan is an interesting one. Psalm 104 contains a reference, yet it is in the context of Leviathan being created by God: 'There go the ships, and Leviathan that you formed to sport in it' (ie. in the sea, verse 26). The sea itself in psalm 104 is simply another part of God's whole creation (see verses 24 and 25), which is celebrated by the psalmist as being created in the wisdom of God. In psalm 148 even the 'sea monsters and all deeps' (verse 7) are called upon to praise the Creator.

Yet by contrast the picture of Rahab and the sea in Isaiah 51 (verses 9 and 10) is anything but positive. This text indeed seems to support Keller's line of argument, because here the prophet is exhorting God to act as in 'days of old' (Is 51:9) when God 'cut Rahab in pieces' and 'dried up the sea, the waters of the great deep'. Yet even here the context of the phrases is important. God dried up the sea, in order that the redeemed might cross over (see also Exodus 14:1–31). Keller neglects to mention that the focus is not on the sea, *per se*, but on the action of God, who redeems God's chosen ones and brings them back to Zion. This short examination of some of the biblical texts containing references to the deep or to the sea monsters simply illustrates that the ancient Jewish understanding of the sea was not uniform.

Whilst Keller's connection between the book of Revelation and the current environmental crisis is to be understood simply as an illustration, the fact that she makes so much of the connection means the focus of her work becomes somewhat clouded at this point. She is at pains to maintain a sense of not becoming premillennial, and yet, if one were to read her work without that proviso, there is a sense in which she slips into the kind of language normally associated with apocalyptic groups. Indeed, as I will suggest in the next chapter on apocalypticism, there is a sense in which she fits into the category of 'negative apocalyptic' (which I will also define in

19. Keller, 'No More Sea', 190.

the next chapter). This may well be a deliberate ploy to provoke a reaction leading to environmental action. If so, she certainly pursues it as a method to full effect. This comes out very strongly in an article published in the journal *Ecotheology*.[20] Here Keller examines in more detail the connection between the prophetic utterances of the book of Revelation and the capitalist approach to the environment. Choosing a number of key phrases, she examines how these prophecies are finding expression in the attitude of society to the environmental crisis. So the utterance, '*because you are lukewarm, and neither cold nor hot, I will spew you out of my mouth!*' becomes an expression of the lukewarm indifference espoused by many, which suits corporate interests. The expressions, '*the kings of the earth have committed fornication*' and '*the merchants of the earth have grown rich with the wealth of her wantonness*' become an indictment of capitalist society, with its imperial power and economic injustice. The mark of the beast, which means '*that no one can buy or sell unless he has the mark*', far from being a prediction of credit card technology, as some fundamentalist groups maintain, becomes a symbol of the almost total captivity of Western societies to the concept of free trade, in which no people 'may resist the international power of the transnational corporations and the flow of their capital'.[21] Keller suggests that the worship of the beast can indeed be likened to the fact that we can hardly even imagine an alternative to the dominant system of world free trade and international markets.

Yet for all her enjoyment in using the book of Revelation, she ultimately rejects its final message, a message of hope. Curiously then, she is willing to use the imagery of apocalyptic destruction to make a point about the world environmental crisis, but is then unwilling to use the imagery of hope to make a point about eschatological hope. This playing of scripture off against itself is an unhelpful theological method. In focusing so intensely on the book of Revelation and in using the apocalyptic language that she does use, she ultimately misses the opportunity to develop a more balanced eschatology, particularly because she rejects the final images of that book. The remainder of the article, far from being a creative reinterpretation of the book of Revelation (which Keller manages in the first part of the article with great success and persuasion), descends into a vague suggestion that we trust in the divine wisdom (the Trinitarian

20. Catherine Keller, 'The Heat Is On: Apocalyptic Rhetoric and Climate Change', in *Ecotheology* 7 (1999): 40–58.

21. Keller, 'The Heat Is On', 54.

Spirit) who will not 'clean up our mess or cool down our planet'[22] but will nevertheless conspire with our best efforts. Thus, according to Keller, John's Apocalypse cannot guide us so that we have strength of purpose or a delight of vision. Indeed she somewhat peremptorily condemns the message of hope contained in the book as:

> inducing pathetic hopes for a quick fix, predicated on an androcentric authoritarian picture of God, a loveless warrior messiah, a chillingly regimented nature serving the architecture of the New Jerusalem, the denial of death and the torture of the damned.[23]

The grand vision of a new heaven and a new earth has no place in Keller's appreciation of the book of Revelation. The book cannot guide us to a sustainable future. Keller sees no long term, applicable hope in its message. Yet she does not want to leave the book in the hands of fundamentalists either. So she does what many people do when reading scriptures: she picks out the bits she likes. She picks out those apocalyptic metaphors helpful to the environmental cause, but abandons the book's main message of hope and consummation.

1.7 Human accountability

Having made the above criticism of her work, it's important to note that the crux of Keller's criticism of traditional eschatology indeed comes when she suggests that human accountability for the earth and its sustenance is enveloped by the hope for resurrection into deathlessness. The possibility of environmental activism becomes moot for the Christian, because all suffering and death will be overcome. Moreover, by a strange twist of Christian apocalyptic logic, the closer we draw to the end of all things, the worse things become, and the closer we are to ultimate transformation. This is precisely the problem with Christian eschatology as it has developed, argues Keller, because there can be no human contribution to the systems of domination and destruction, only a supernatural solution. Thus prolonged human activity in the field of environmentalism becomes completely pointless because, 'human accountability for the polysystemic

22. Keller, 'The Heat Is On', 57.
23. Keller, 'The Heat Is On', 56.

ecosocial injustices of the earth is swallowed by the ultimate hope for resurrection into deathlessness'.[24]

There's no doubt that Keller is motivated out of a deep concern for the environment when she is highly critical of the system of eschatology as presented in the classic biblical eschatological texts. What she utterly fails to demonstrate is the capacity to reinterpret traditional eschatological models to take account of environmental concerns, particularly the question of complexity and diversity. The consequence of her failure to do this is a model of eschatology which offers no ultimate hope and is of limited pastoral use. She does include a kind of vague suggestion that the species lives on in its children. But even this hope is fundamentally flawed, because the earth as we know it is ultimately doomed to destruction. In the end the universe will cease to exist, at least in any recognisable form, were there any intelligent life left to observe it.

Keller simply does not acknowledge this. Instead she calls for a model of eschatology based on a number of concepts arising principally from chaos theory. What is missing from her eschatology, somewhat strangely, is any development of an understanding of God. Instead she relies on scientific theories. Drawing theological conclusions from scientific theories is a theological method potentially rich in symbolism, but it must surely be balanced by equally well developed concepts of God and especially God's providence. Keller fails to do this. Whilst she argues that her use of the concepts are figurative, not foundational, in fact they become foundational because of a lack of other developed theology, certainly in her article 'No More Sea'. The first concept is that of *non-predictability*. Because the universe seems to be genuinely creative, we are released from any concept of determinism or of eschatology as some kind of guaranteed outcome. Closely related to this point is the assertion, however, that *interrelatedness* is the foundation of all ecosystems. Because of this interrelatedness, and coupled with the non-predictability, there is hope for the universe because a relatively minor movement (ecological Christianity for example) might influence the processes of the universe disproportionately. Because of the interweaving of order and chaos, of self-similarity and difference (for example in fractal geometry) there is a complexity to creation, in which one part of creation is never kept outside of another, but forms a complex inextricable web. Thus, using Keller's example, *tehom* and *eschatos* cannot be separated. These theories point to a concept of creation at the 'edge of

24. Keller, 'No More Sea', 187.

chaos', which is the evolutionary site for 'complex self-organising systems'. These systems, where there is 'enough order and enough chaos for what is novel, interesting, creative and complex to take place',[25] will provide theology with the basis for an eschatology, which is 'tehomophilic'. In other words, the complex interplay of order and chaos, where neither one nor the other gains an ascendancy, is the place theologians may express their hope and so develop an ecological eschatology. Keller suggests that the sustaining of a complex and bio-diverse creation provokes a bottomless love, which is itself its own reward and needs no final prize. No simplicity will save us, she suggests, rather it is the rhythm of life, by which we again and again centre ourselves and in which we are embodied, which becomes our home. There is no need for a new heaven or a new earth, because we have this Earth, this sky and water to renew.[26]

1.8 Conclusion

There are elements of Keller's theology which are deeply persuasive. Certainly her concern for the long term effects and consequences of the environmental impact of the human species are shared by many theologians. There *is* a need to respond to the problems of *this* world,[27] rather than simply wait for a fiery end and the creation of a new habitat. We do indeed need to encourage and further develop a Christian spirituality, which imagines a long and sustainable future on this planet, which denounces injustices perpetrated against the poor and oppressed, including creation itself, and which rejoices in the complexity of life and the diversity of culture, language and the rich diversity of life around us. There is indeed in the Christian tradition much 'radical wisdom',[28] as Keller calls it, which needs to be recycled. She points to the eschatological energies of the prophets (including in this description Jesus and Mary[29]) for whom prophecy meant a 'denunciation of injustices against the vulnerable and

25. Keller, 'No More Sea', 195.
26. Keller, 'Women against Wasting the World', 263.
27. For a good summary of this position see Kosuke Koyama, 'Cosmology and Justice in Ecumenical Perspective', in *Theology for Earth Community*, edited by D Hessel (Maryknoll, NY: Orbis, 1996), 103–112. Koyama suggests there is no 'getting out' as she puts it, but rather, 'the biosphere is the primary location of human salvation'. Koyama, 108–109.
28. Keller, 'Women against Wasting the World', 261.
29. It's important to note that only Islam specifically refers to Jesus as a prophet. Traditionally, neither Christianity nor Islam considers Mary a prophet.

the vision of a lush future in *this* world for all who partake in justice and wisdom'.[30]

Keller does acknowledge, however, that many of the prophetic texts, and here she uses Mary's *Magnificat* as an example, do express the hope that the fullest realisation of the prophetic vision lies in the future. She is concerned that a deliteralised and deapocalypticised eschatology is needed, if the feminist project of ecocentrism is to be served in a socially and historically responsible manner. Herein lies a fundamental problem with Keller's work and indeed that of those feminist writers who have rejected an eschatological hope. Their work and the Christian tradition begin to part ways. A brief examination of some of the parables of Luke's gospel will serve as an example. The Jesus portrayed in the gospels was passionately concerned about justice for the poor, the oppressed, and the outcasts of society. His concern was immediate. He seems to have wanted immediate justice. But many of his parables (*at least as developed and presented by the gospel writers—this is an important qualification*) also show a sense of eschatological justice as well. Moreover, this sense of God's eternal justice functions in the parables as a method for causing justice to be enacted in *this* world. In other words, the threat of God's retribution on those who practice injustice functions as an incentive to practice justice now.

The parable of the rich man and Lazarus is a good example (Lk 16:19–31). Because the rich man completely ignores Lazarus during his life-time, despite Lazarus' poverty and his own wealth, the rich man is punished eternally. Another example is Jesus' promise that those who are generous now will be rewarded at the resurrection of the righteous (Lk 14:12–14). Jesus encourages his listeners to invite those people to dinner who will not be able to repay them (the poor, the crippled, the lame, and the blind), because the repayment will come at the resurrection. The apocalyptic sayings also contain this element of the coming judgment, in which our lives in *this* world will be judged by God (see, for example, the sayings about watchful slaves: Luke 12:35–48). All of these parables function not only as teachings about the coming judgment of God, but also as incentives for people to live lives worthy of Christ, beginning in the here and now. But Keller abandons this tradition completely. Because of what she leaves out of the Christian faith Keller loses a significant and important part of its history. She also loses a potential ally in the tradition for environmental action. I will develop this line of argument below, but, briefly, if we ac-

30. Keller, 'Women against Wasting the World', 261.

knowledge that God will judge us (however fairly and lovingly) for our actions in this life, we are motivated to live lives which are environmentally and socially responsible.

Aside from that particular hermeneutical criticism, Keller's arguments might indeed be convincing, if they were taken in isolation. They are fundamentally flawed, however, because creation itself is temporal and transient. The cosmos is not timeless, nor is it renewable in perpetuity. Rather there is an inevitable destruction which waits for all of creation, beginning with Earth itself. The second law of thermodynamics and the predictions about the continued expansion of the universe tell us this. It is simply untrue to assert, as Keller does, that 'the deep, the indeterminate potentiality in which we swim, which will hold the seeds of future earth evolutions no matter what we do, does not end'.[31] The terrible truth, which faces us all at our own death, but which is true for the whole created order as well, is that it *will* all end. Whether that end will be the final word is the question which eschatology has sought to address. But there is no escaping the reality of the end of the universe as we know it, except through the participation of the divine. It is only when we begin to discuss God's involvement in the divine transformation of all things, that a picture begins to emerge, which is different from the one scientists present. Without God, without divine participation, there is no hope for the universe. That much is determined already. In that sense, her work on the unpredictable and indeterminate universe, this complex and inextricable web of relationships, is a useful model for a doctrine of creation, but limited in its use for eschatology, because there is no development of a *theology* to support the model.

The criticism of Keller as her work pertains to pastoral care is a relatively undeveloped one. In her response to Keller's article 'No More Sea: The Lost Chaos of the Eschaton' Mary Hinsdale suggests that 'Keller's ecological eschatology would seem to require a relinquishment that may be fraught with emotion for many Christians'.[32] Hinsdale refers in passing to the fact that the particular reading from Revelation 21:1–4, which Keller critiques so strongly, was the one chosen by her family for her (Hinsdale's) father's funeral. What Hinsdale alludes to, but doesn't develop, is that Keller leaves us with a version of Christianity, which is not only divorced from the eschatology of the New Testament, but also in crisis in a pastoral sense. What is the pastoral approach to the bereaved, needing to be

31. Keller, 'Women against Wasting the World', 261.
32. Mary Ann Hinsdale, 'Response to Catherine Keller', in *Christianity and Ecology: Seeking the Well Being of Earth and Humans*, 200.

practiced by priests, ministers and pastors as well as trained lay people all over the world every day, if not one which includes the hope of the resurrection? Certainly there is a balance to be found between, on the one hand, acknowledging the death of a loved one and permitting, even encouraging, the grieving process for that loved one, and, on the other hand, sustaining individuals and families with the hope of the resurrection. It is a balance sought by priests and ministers each time they visit the bereaved in preparation for a funeral service. It is this very hope which, as Hinsdale points out, Keller finds problematic. Yet she offers no viable alternative, in my opinion, because her vision of us centring ourselves in the rhythm of this world, this earth, this sky and this water,[33] is of little comfort in an ultimate sense.[34] There is no redemption or justice for those cut short in the prime of their lives, there is no comfort for the bereaved, and, ultimately, there is no hope, because this world will eventually come to an end, even if we manage to avert the current ecological crisis and continue to avoid nuclear war.

In abandoning the Christian concept of the resurrection of all things, including not just the resurrection of human beings but also the transformation of creation itself, Keller is in fact abandoning a significant factor in environmental ethics. The very ethics which she seeks to incorporate into theology are at risk. If the scientific predictions of physicists about the end of the world are at all accurate, then the earth and indeed universe as we know it is headed for destruction. There is no ultimate hope in the processes of the world, at least not by themselves in the way which Keller suggests. So, the question must be asked, if creation as we know it is not destined for resurrection, but rather ultimate destruction, why bother caring for it? Indubitably Keller's theology allows for other reasons for being concerned with environmental ethics than simply the promise of the resurrection. However, in abandoning traditional eschatology, she is in fact losing a potentially powerful ally in the drive for environmental action. The ecological eschatology, which I hope to develop in this work, suggests that the environmental actions we take *now* are of eternal significance. What we do now, in *this* life and to *this* world, matters eternally to God. Just as the complexity and beauty of the evolutionary processes of life will be gathered up into God, remembered and treasured, so too the current environmental state of the world and the actions we take to preserve it,

33. Keller, 'Women Against Wasting the World', 263.

34. However, I acknowledge that many people (such as green activists, for example) do find some comfort in this approach.

will be remembered by God eternally. To lose the hope of the resurrection, particularly the hope of the transformation of all things, is to lose a significant impetus for environmental ethics and therefore environmental action.

Chapter Two
Rosemary Radford Ruether

2.1 Introduction

Turning to the work of Rosemary Radford Ruether we find similar problems to those afflicting Catherine Keller's work. Like Keller, Ruether seeks to deconstruct much of classical theology from a feminist perspective. Like Keller she also successfully identifies many of the shortcomings and underlying assumptions afflicting classical theology. And, without preempting my analysis, Ruether, like Keller, fails to reconstruct a Christian eschatology, leading her readers instead to a position one small step removed from pantheism, though she vigorously defends her work against such an accusation.

2.2 Creatio ex nihilo

Ruether's deconstruction of classical theology, at least in terms of the relevance of her work to Christian eschatology, begins with her analysis of the Christian doctrine of creation. She describes the Christian doctrine of creation as a synthesis of a number of ancient ideas from different cultures, namely, the Hebrew Scriptures and Greek (Platonic) as well as other ancient Near Eastern cultures. This results in an ambivalent attitude to creation, which Ruether identifies as one of the root causes of all the problems. The Christian concept of an eternal, absolute sovereign God, existing since before the beginning of time, whose sovereignty cannot be challenged, meant that the Hebrew and Platonic understandings of creation coming from a pre-existent chaos were unacceptable. The Christian doctrine of *creatio ex nihilo* thus began to be important, because the stuff of matter could not somehow have been there in the beginning—God created the matter itself, before forming it into the cosmos. But this doctrine of *creatio ex nihilo* meant that Christian theology as it developed was left with an unresolved ambiguity towards creation. If it comes from God then in some way it has emanated from the divine being, yet there must at all times be a fundamental distinction between God and the creation.[1] This ambiguity towards creation found expression in the doctrine of God, which wanted to assert, on the one hand, that the Word and the Spirit were present in creation, 'grounding its existence and being revealed in history, and finally bringing creation itself to transformed communion with God', [2] and yet on the other hand, that the second and third persons of the Trinity are fundamentally distinct from creation. So the Christian doctrine of God was faced with the problem of expressing how God was fundamentally distinct from creation and yet, in some sense, was intimately involved in creation, because creation was seen as 'welling up out of and being sustained in existence through the being of God'.[3]

Ruether's key point here is that the doctrine of *creatio ex nihilo* 'leaves Christianity with an unresolved ambiguity about the ontological status of "matter."'[4] It comes from God, and yet is fundamentally distinct from God. This question of the ontological status of creation and its ambiguity, from Ruether's point of view, in the Christian tradition forms a fundamental basis of her work. It is precisely because of this perceived ambiguity towards the ontological status of matter that Christianity has exploited creation and contributed to the current environmental crisis.

Having asserted that the ontological status of matter was ambiguous almost from the very beginnings of Christianity, Ruether then develops a theology of eco-feminism based on the concept of Gaia. However, a closer examination of the early Christian debates on the doctrine of *creatio ex nihilo* calls into question her assertions. Certainly the doctrine of creation became a vexed question for Christians, particularly in the second century. However, it was precisely in response firstly to neo-Platonism and then Gnosticism, and indeed in strong reaction to them, that the doctrine of *creatio ex nihilo* was developed. It was to resolve the very ontological questions that Ruether is raising that a number of Christian apologists of the second century began to debate the issue in earnest. Unlike the Jews and Christians of the first century, the Christian theologians of the second century began to think more precisely about the ontological status of

1. Rosemary Radford Ruether, *Gaia & God: An Ecofeminist Theology of Earth Healing* (London: SCM, 1993), 27.
2. Ruether, *Gaia & God*, 27.
3. Ruether, *Gaia & God*, 27.
4. Ruether, *Gaia & God*, 26.

creation. The traditional saying, that God created the world out of nothing, took on 'a principled ontological sense: the expression "out of nothing" now meant that absolutely, and excluded the idea that the creator has merely imposed form on a pre-existent material'.[5]

This was a significant development in Christian doctrine from Christianity's Jewish roots. Whilst there are references within ancient Jewish literature to a doctrine of creation out of nothing, they are isolated statements, arising from ongoing discussions. They were not developed into a formulated, unambiguous doctrine. Whilst the absolute creative activity of God is defended when attacked, a detailed doctrine was not developed, at least not in the way that it was relatively quickly developed in the life of the Christian movement.

Gerhard May discusses some of the reasons why ancient Judaism was comparatively unconcerned by such questions. Certainly Christianity's collision with Gnosticism meant that it was forced to develop a coherent theology of creation, partly to protect the concept of divine omnipotence: God must be distinct from creation. But Jewish thought seems not to have been directly affected by the rival philosophical movements on the question of creation out of nothing until the Middle Ages (*see below*). Gerhard May asserts that this has nothing to do with Jewish thought being somehow unspeculative—the historical effectiveness of both Jewish and Christian thought indeed rested on the fact that they could become philosophical. Rather, to rabbinic Judaism the questions raised by Greek ontology were relatively remote. Critically, a specific doctrine of *creatio ex nihilo* was not necessarily demanded by the text of Genesis. The mention of chaos in the opening verse of Genesis could also support the view 'that an eternal material existed, which God had merely ordered in creating the world'.[6] Importantly too, May asserts, Jewish thought 'is in its entire essence undogmatic'.[7] At least in the question of creation, Jewish thought did not find itself tied down by statements in the Bible. It was not until the Middle Ages that Jewish philosophers, in debate with Arabic neo-Platonism and Aristotelianism, developed a specific doctrine of *creatio ex nihilo*, but even this was not seen as the sole valid position.[8] This arguably more flexible approach has its own strengths (in allowing more

5. Gerhard May, *Creatio Ex Nihilo: The Doctrine of 'Creation out of Nothing' in Early Christian Thought* (Edinburgh: T&T Clark, 1994), 22.
6. May, *Creatio Ex Nihilo*, 24.
7. May, *Creatio Ex Nihilo*, 24.
8. May, *Creatio Ex Nihilo*, 25.

speculative theology, for example) but what it demonstrates is the highly developed nature of the Christian doctrine of creation out of nothing by comparison.

The Christian position had a different starting point from the Jewish faith. It began with the person of Jesus Christ: the Easter faith, specifically, forms the starting point for all Christian dogma. This is a significant difference, one which had the practical consequence that Christianity had to deal with philosophy in a decidedly more critical way than Judaism, despite Judaism's being in a similar 'missionary situation',[9] as Gerhard May puts it. Developing a hermeneutic of the Christ event, which is *the* salvation event in the Christian paradigm, in turn influenced the Christian understanding of creation. Not only is the Christ event understood as a creative act of God, but the relationship of the Second Person of the Trinity to the rest of the creation has to be resolved. It is here, in the discussion of Christology, that critical differences had to be shown to exist between neo-Platonist cosmology and Christology. All of creation had to be seen as dependent upon the Second Person of the Trinity, who stands at God's side. There could be no sense of creation somehow having existed eternally, because then the place of the Second Person of the Trinity in the processes of creation would be compromised: God the Son would simply be the shaper of pre-existing, timeless matter, like God the Father would be in this paradigm. Early Christianity was concerned to preach that the Second Person of the Trinity was intimately involved in the creation of the world, with the other persons of the Trinity, and that the world continued to exist because it was upheld by God, with Christ having ascended to be with God the Father.

It is significant to note, however, that although the intention of the New Testament statements about creation is 'quite legitimately interchangeable with the idea of *creatio ex nihilo*',[10] that particular cosmological theory only developed in all its subtlety when Christians encountered firstly neo-Platonism and then Gnosticism. Crucially, though, creation is not isolated from God's salvific activity. It is a fundamental part of it. Indeed the act of creation is the beginning of God's salvific action.[11] There was no distinction made between God's activity in creating the world and his beginning of salvation history—the two are intertwined. This is a critical point, because the salvation of human beings does not occur in isolation from the

9. May, *Creatio Ex Nihilo*, 25.
10. May, *Creatio Ex Nihilo*, 26.
11. See May, *Creatio Ex Nihilo*, 28–30.

rest of creation. Indeed the whole of creation, not just human beings, was seen as being 'dependent on Christ and subordinate to him, who himself stands wholly at God's side'.[12]

In the second half of the second century, the early Christian doctrine of creation developed significantly, as stated above, when Christianity increasingly found itself confronted by neo-Platonism and Gnosticism. Tatian, Theophilus of Antioch and particularly Irenaeus all helped define the Christian doctrine of creation over and against the rival philosophical views. In Irenaeus the themes of creation and salvation are critically connected.[13] Creation is contingent upon God's grace. Indeed the very ground of creation is God's goodness; God's sustaining of creation is a free gift. God holds this transitory world in being through his will and leads all creatures on to perfection. This holding the world in being is not some kind of static behaviour, rather God is continually involved in saving acts, which reach a climax in the Incarnation of the Logos, 'through whom in the beginning he had created the world and man (*sic*)'.[14] The link between creation and salvation is made explicit, because it is only in connection with the universal economy of salvation that the meaning and nature of creation are understood.

The key question is then, whether there was indeed some unresolved and ambiguous attitude to creation, or whether the Christian doctrine of creation and God's relationship with it was in fact rather carefully and painstakingly developed. Ruether asserts the first position, but an examination of the debates of the first two centuries of Christianity doesn't support her assertions. What we see in the debates about the doctrine of creation between Christianity and its philosophical rivals is the development of a detailed and rich theology of creation. The debates of the second century and the subsequent development of a specifically Christian doctrine of creation are therefore considerably more subtle than Ruether allows for in her opening chapters of *Gaia and God*.

To be fair, Ruether does briefly acknowledge some of the complexity in her discussion of Irenaeus. Curiously though, in acknowledging the contribution of Irenaeus, Ruether almost seems to contradict her earlier assertions about the development of the Christian faith. She quotes Ire-

12. May, *Creatio Ex Nihilo*, 29.
13. See also Brian Daley, *The Hope of the Early Church: A Handbook of Patristic Eschatology* (Cambridge: Cambridge University Press, 1991), especially chapter 3 'Regaining the light: eschatology in the Gnostic crisis (150–200)', 25–32.
14. May, *Creatio Ex Nihilo*, 176.

naeus as an example of an early Christian philosopher with a wide vision of the consummation of all things, for whom 'creation is itself an incarnation of the Word and Spirit of God, as the ontological ground of bodily existence'.[15] Thus the earthly or bodily nature becomes the 'sacramental bearer of the divine, and the divine deifies the bodily'.[16] But her acknowledgement of Irenaeus' theology would seem to contradict her earlier dismissal of Christianity as having an unresolved ambiguity about matter's ontological status.[17] Irenaeus can be seen as the climax of a developing understanding of the Christian doctrine of creation out of nothing, in response to the philosophical challenges of the day. This doctrine of *creatio ex nihilo* becomes the accepted Christian position even for Tertullian and Hippolytus. Irenaeus was a key thinker in helping focus Christian thinking on the whole question of the status of matter.

Ruether's examination of Irenaeus is telling, because although she acknowledges the legitimacy and strength of his theology, her own theology heads in a completely different direction. She ignores the opportunity to explore some of Irenaeus' concepts of the consummation of all things, focussing in a rather more narrow way on her own particular and limited view of eschatology. Ruether's development of theological concepts seems incapable of including any concept of eternity. At every turn it is limited by the fear that theology might dare to include some concept of human beings or creation itself transcending mortality. Thus even when examining Irenaeus' vision of the whole cosmos being united with the divine life of God, Ruether stops short of accepting this as a possible current model of eschatological fulfilment.

It is to Ruether's criticism of the desire to transcend or 'escape' mortality that I now turn, in the examination of the Christian understanding of 'body and soul' and then apocalypticism.

2.3 Body and soul

Leaving aside the particular historical questions surrounding the doctrine of creation, we come now to the question of the influence of platonic concepts of body and soul on Christianity. Ruether goes on to assert that the unresolved ambiguity towards creation finds potentially its most destructive expression in the Christian understanding of the body and the

15. Ruether, *Gaia & God*, 235.
16. Ruether, *Gaia & God*, 235.
17. Ruether, *Gaia & God*, 26.

soul. Unfortunately, from Ruether's perspective at least, Christianity was heavily influenced by Platonism, which means that many of the Christian concepts and doctrines are inaccurate reflections of the Hebrew understanding of the body. The body-soul dualism in human beings reflected the dualism of the God-cosmos division in Christian thought. Influenced by Platonism, the Christian position reflected an hierarchical understanding of the cosmos. The earth is seen as occupying the lowest possible position in the hierarchical levels of the cosmos. The earth, and the region between the earth and the moon, is seen as the realm of mortality, whilst the supralunar region and the fixed stars are seen as being composed of immortal matter. Beyond the stars is eternal space, where God dwells with the angels and saints. According to Ruether, this world view, which was a meshing of the Platonic picture of the world with Christian eschatology, was unchallenged until the seventeenth century. The Christian concept of the human soul reflected a synthesis of Greek and Hebrew ideas, heavily influenced by this Platonic hierarchical and dualistic cosmic framework.

Platonism believed that the human soul was pre-existent, originally dwelling in the stars in the upper regions of the cosmos, before being incarnated. Hebrew thought, in marked contrast, believed that the soul was the life principle of the body, 'and future life takes the form of a resurrection of the body on a renovated earth'.[18] The Christian position, particularly as it developed at the hands of Augustine, attempted various syntheses of these two positions. The dominant Augustinian model (a development of St. Paul's writings, which Ruether also examines) considered that human beings were immortal prior to sin. However, having been created with immortal, spiritual bodies, they fell into mortal forms of bodiliness with sin. The orthodox Christian position—as opposed to those parts of Christianity heavily influenced by Gnosticism—rejected the concept of a pre-existent soul and also the concept of reincarnation. Thus 'each soul is created uniquely by God and with equal capacity for holiness and redemption'.[19]

The orthodox Christian position as it developed, however, displayed a number of inconsistencies or ambiguities. For example, the position of men and women in society did not reflect the equality of souls in relation to God. Whilst men and women were equal before God in redemption, they were not so in this unredeemed world. Female subordination to the

18. Ruether, *Gaia & God*, 28.
19. Ruether, *Gaia & God*, 28.

male was understood as being both natural, reflecting an inferior body and personality, and the punishment for causing original sin:

> Thus God seems to will different principles in creation and redemption. In creation women are inferior and under male subordination, yet this inferiority is abolished in their equal capacity to transcend creation and be transformed by grace.[20]

This inequality between male and female reflects the Christian framework more generally, in which there is a distinction made between the body and the soul. Whilst the Christian tradition, particularly Latin Christianity, accepted the Hebraic view of the concept of the soul as the life principle of the body, it also accepted the Platonic understanding of the soul being capable of being detached from the body. The body-soul relation thus became ambiguous, because the soul was thought to exist in a disembodied form after death, though it would be rejoined with a spiritual body at the resurrection, which is when the final eschatological fulfilment comes for all things: the final consummation of all things.

Critically, this concept of the created soul, which is nevertheless capable of an immortal, transcendent life apart from the body, made clear the distinction between human beings and the rest of creation. Human beings, with their rational soul, are destined for immortal life, but the lower orders of being, animals and plants, have a lower life and lack this higher soul, which is rational and immortal. As a consequence the immortal soul will (presumably at the time of death) kick aside the physical world of bodily life and ascend to its immortal life, escaping from and unencumbered by the limited life and mortal fate of the rest of earth's creatures. Whilst Ruether acknowledges that the official doctrine of Christianity affirmed the concept of the resurrected body and a redeemed earth, she suggests that the operative Christian eschatology was this Platonic notion of the immortal soul ascending to paradise.

Ruether argues that a direct consequence of this particular Platonic understanding of the nature of human beings and other creatures is that human beings have no ultimate responsibility towards the rest of creation. Again, this is the result of a somewhat ambiguous Christian picture of the world. The Christian concept of the world includes the notion that originally the world was perfect. This is reflected in the first account of

20. Ruether, *Gaia & God*, 29.

creation in the book of Genesis: the expression 'very good' is used, as Ruether points out.[21] The original earth was paradisal. Yet somehow evil entered into the world. Natural evil, that is, the death and violence associated with creation itself, including in the animal kingdom, was included in the concept of evil. The concept of evil, however, also included the violence of human beings towards one another. Human sin was considered to be the cause of *all* evil, not just the violence of human beings towards one another. Through sin comes death, not just to human beings, but to the whole created order. Human beings are rescued from this state of being through the sacrifice of Jesus Christ, yet this leaves the state of the rest of the cosmos in some doubt. The classical Christian solution was then to see the whole cosmos as eventually being redeemed, although Ruether suggests that this only applied to planetary structures and not to plant and animal life.[22]

Unfortunately Ruether gives no reasoning or evidence to back up this last assertion, so it's somewhat difficult to refute it. However, she does remind us, that the *Hebraic* understanding of 'creation' and 'redemption' did not involve a separation of the two concepts—they were inextricably connected.[23] The welfare of the land, including its fruitfulness and the wellbeing of the animals, were connected with the faithfulness of the people of Israel and God's blessing. This connection is particularly evident, Ruether argues, when the sabbatical legislation is examined. She considers this legislation the 'fruit' of Israel's belief that God's blessings were connected with Israel's faithfulness. The cycles of life, based around the number seven, namely seven days, seven years, and seven-times-seven years or Jubilee, assured the people, the land itself and the animals of a time of rest and restoration. These cycles were the practical expression of the connection between justice and God's blessing. This Hebraic vision, particularly the Jubilee year in which there was meant to be a 'yet more profound renovation of relationships between humans, animals, and land',[24] recognised that human beings periodically drifted into unjust relations with one another, the animals and the land.

21. Ruether, *Gaia & God*, 30.

22. Ruether, *Gaia & God*, 30.

23. Ruether, *Gaia & God*, 207–214.

24. Ruether, *Gaia & God*, 212. Whether the Jubilee year ever took place in history is a different question altogether.

Ruether then suggests that the Hebraic vision of 'periodic righting of relationships' *did* involve a 'more absolute messianic future'.[25] She cites the prophetic vision of Isaiah 65 as an example of this, in which peace between people will be finally fulfilled and nature's enmity healed. She points out that even the 'carnivorous conflict between animals will be overcome in the Peaceable Kingdom'.[26] However, she insists that this vision of a messianic future is 'still a future within history'.[27] The example is contrasted with what she considers the Christian spiritualisation of such worldly concerns for justice and the reinterpretation of the idea of the promised land as the whole cosmos, 'and particularly as that part of the cosmos transcendent to earth, or "heaven"'.[28]

However, Ruether's interpretation of Isaiah 65:17–25 as a hope for the *historical* fulfilment of Sabbath peace is not the only way to interpret the passage. Indeed Carol Dempsey suggests that the visions of the prophets, including the vision of Isaiah 65, 'are not yet fulfilled because even though they are part of a particular historical time frame and relate to contemporary situations they *are* eschatological'.[29]

According to Ruether the consequence of the concept of bodily resurrection for human beings is an ambivalent attitude to the rest of creation. Human beings, by reason of our sin, are said to be responsible for the shortcomings of the rest of creation. Yet, on the other hand, we bear no responsibility for the rest of creation. Plants and animals can be exploited at will. They have no personhood that needs to be respected. And, asserts

25. Ruether, *Gaia & God*, 213.
26. Ruether, *Gaia & God*, 213.
27. Ruether, *Gaia & God*, 213.
28. Ruether, *Gaia & God*, 215.
29. Carol Dempsey, 'Hope Amidst Crisis: A Prophetic Vision of Cosmic Redemption', in C Dempsey and R Butkus edited, *All Creation is Groaning: An interdisciplinary Vision for Life in a Sacred Universe* (Collegeville, Minnesota: The Liturgical Press, 1999), 280. Dempsey's article provides an eschatological perspective on the interpretation of several of the visions in the books of Isaiah, Jeremiah, Ezekiel, Hosea, Joel and Amos. See, however, Claus Westermann, *Isaiah 40–66: A Commentary* (London: SCM, 1969), 298–299 and 408–409. Westermann argues that Isaiah 65:17a and 25 were quite possibly added later to Isaiah 65:16b–25 and reflect a later apocalyptic concept. He suggests that 'the salvation which Trito–Isaiah proclaims is conceived in terms of this world. That is to say, the great change which inaugurates it brings about a state of affairs which is still tied to the realm of history.' Westermann, *Isaiah 40–66*, 298. The few passages which go beyond this, including Isaiah 65:17a and 25, 'where the salvation promised has traits incompatible with the realm of history' are not characteristic of Trito–Isaiah's own proclamation. Westermann, *Isaiah 40–66*, 298–299.

Ruether, in the classical tradition, they do not share in our eternal inheritance.

One possibility in the development of Ruether's work, of course, would be to explore how plants and animals might indeed be included in an understanding of resurrection or eschatological fulfilment. This is part of the focus of this book, but Ruether's work moves in an entirely different direction. Her work moves towards an understanding of a common organic mortality, shared by human beings, plants and animals alike. Any reshaping of nature must reflect an ecological ethic based on the impulse to consciousness and kindness. We do not pass on our ideals to the future by escaping personal death, but by this reshaping of nature, which is nevertheless 'finally governed by the finite limits of the interdependence of all life in the living system that is Gaia'.[30]

As I shall discuss below, this qualification by Ruether, that any hope for the future works within finite limits, in fact places a limit on God. Ruether's theology limits what God can do. The possibility of eschatological fulfilment, including the possibility that all of creation (plants and animals as well as the planetary structure) is part of that fulfilment, is rejected and said to be impossible. Ruether restricts God's potential activity and confines her understanding of God to that of an earth goddess, intimately bound up with creation and incapable of doing or creating something genuinely new and original. The fundamental flaw in Ruether's theology, as I shall discuss in more detail at the end of this chapter, is *theological* rather than simply eschatological.

2.4 Apocalypticism

Ruether's placing of limitations on God's potential activity seems to arise from her total rejection of the apocalyptic tradition within Christianity. Her theology as it develops is a reaction against this tradition, with the result that any possibility of a reinterpretation of the tradition or a development of some of its themes, is discarded in favour of a radically new approach to eschatology, ironically something she denies as being possible for God in terms of a new creation. It is therefore in her rejection of apocalypticism that we find the beginnings of her approach to eschatology.

Ruether suggests that early Jewish apocalyptic visions have a quite different focus from the later preoccupation with overcoming mortality

30. Ruether, *Gaia & God*, 31.

(which then became Christianity's key focus). Rather it is in response to repentance by the chosen people of Israel that God restores the fortunes of the land, bringing with that restoration abundant rain, bountiful harvests, the restoration of cities, the domination of enemies and, in the more universalist visions, the conversion of all nations and the establishment of peace and justice throughout the world. The threats of divine retribution are in response to Israel's failure to obey the commandments and her turning away from the Lord. The acts of divine destruction (manifested by conquering armies, or natural disasters such as drought and flood) are, however, not terminal, but rather act as warnings, calling Israel to repentance. After a period of Israel languishing beneath a reign of evil, brought upon it by its own sins and rejection of God, God will bring the reign of evil to an end. The vision then becomes a sweeping historical vision, in which God eventually judges and destroys the wicked nations of the earth 'and vindicates the righteous in a renovated earth, where peace and good times will prevail'.[31]

Ruether asserts that these Jewish apocalyptic visions were probably influenced by Persian religious ideas, particularly Zoroastrianism, in which there is a progressive triumph of good over evil in the fourth period of the history of the world (each period being divided up into 1000 years in the theology of Zoroastrianism). At the end of the history of the world the righteous would be separated from the unrighteous, the cosmos itself transformed and purged of both evil and mortality, becoming immortal and everlasting.

In Hebrew thought this type of apocalypticism finds clear expression in the book of Daniel. However, Ruether maintains that, despite references in the book (for example 12:3–4) to the resurrection of past dead, the key function of resurrection in this early Jewish apocalyptic work is to establish justice and righteousness, 'not to overcome human mortality'.[32] It is only in the later apocalypses of the first centuries BCE and CE that there is a new focus on immortality.

It is Christian apocalyptic thought, which Ruether finds so problematic. She maintains that 'apocalypticism, like Platonic eschatology, is based on the fantasy of escape from mortality. Death itself is the "last enemy" to be overcome.'[33] Crucial to this 'escape', as she terms it, is the destruction of the earth itself, which is associated with decay and death. Through the

31. Ruether, *Gaia & God*, 67.
32. Ruether, *Gaia & God*, 71.
33. Ruether, *Gaia & God*, 83.

destruction of the earth, and the elimination of evil, embodied by those 'enemies' of good, the final victory over death can be won. The 'enemies', Ruether asserts, are often defined by apocalypticists as people belonging to other racial, ethnic, and religious groups. Thus God secures a final victory of good over evil. The root of this model of understanding the Christian's relationship to the world (or lack of it) is an understanding of God, who is completely transcendent and un-relational and who will save the chosen ones from the earth, which is destined for destruction. The apocalypticist is therefore confident of his or her survival in the presence of this transcendent, un-relational God and is indeed not only unconcerned by the possibility of world destruction, whether through environmental or nuclear catastrophes, but is immensely cheered by them. The apocalypticist may in fact take comfort in the persecution of the righteous, whilst considering violence directed against others (those with whom they do not associate) to be evidence of divine punishment. Both forms of violence may be considered signs that the 'last things' are about to take place. Ruether believes that 'the apocalypticist may even oppose efforts to ameliorate poverty, prevent war, or clean up ecological damage, for this is to oppose God's will and retard the final deliverance'.[34] This alleged position of some apocalypticists is a somewhat alarming one, to say the least! Whether it is inevitably indicative of most or all apocalyptic world-views is a question I explore in a later chapter.

Unfortunately Ruether's brief discussion of Christian apocalyptic thought seems to be based on extremist positions and there seems to be no examination of academically respected, contemporary writers of apocalyptic theology. This is an unfortunate but recurring element of her book *Gaia & God*. Whilst I do not necessarily agree with all of her conclusions about various historical theological positions, she does examine Christian history quite extensively. The same cannot be said of her examination of contemporary theology, which lacks any reference to key recent writers in apocalyptic theology or eschatology. Nevertheless Ruether's study of two key historical apocalypticists is helpful. She cites both the historical example of an English Puritan in Mary Cary writing her interpretation of the events in England of 1640–1660 (the civil war and the reign of Parliament and then Oliver Cromwell) as well as the American Adventist Ellen White. Mary Cary's various writings are an interesting study in radical apocalypticism, in which particularly historical figures and events are identified

34. Ruether, *Gaia & God*, 84.

with biblical characters and events in the book of Revelation and Daniel. Not surprisingly, given the anti-royal, anti-establishment flavour of the Puritan movement in general in England at the time, the overthrow of King Charles I is viewed as an indication that the time of affliction has ended and that the rule of the saints is about to begin. England is seen as God's elect nation, where the saints will rule peacefully for a millennium, before a final battle with Satan. This will all end with 'the final judgement, the destruction of the mortal heaven and earth, followed by its immortal recreation.'[35] Ruether comments, 'Thus did one English prophetess confidently survey the future from the years immediately after the execution of King Charles by the Puritan army.'[36]

Across the Atlantic, American Protestants were the inheritors of the apocalyptic tradition, with the significant difference that America was seen to be God's elect nation and promised land. There are arguably still very strong elements of this in American popular culture. The apocalyptic elements within American culture owe much to Ellen White's *The Great Controversy*, which continues to be a foundational text for apocalyptic groups within America. Indeed the work was reprinted in 1988 and is still distributed in America under the title *America in Prophecy*. Ellen White belonged to a core group of leaders in a small remnant of the Millerite movement. The movement was named after William Miller, a Baptist minister who preached that the end of the world would come in 1843.[37] Ellen White was accepted by the remnant of the movement as a prophetess. Her ideas have become the basis of American fundamentalist apocalypticists. The critical concept in her work for the purposes of Ruether's discussion is the vision that the 'last things' will begin with the persecution of the righteous, but that Christ will return when the persecution reaches its high point, and save the righteous. 'The righteous will be taken up into heaven to enjoy a thousand years of felicity. Meanwhile the earth will be handed over to desolation and destruction. It will be utterly laid waste (drawing on Isaiah 24).'[38] After a thousand years of Satan's rule of a desolate earth, a final battle will see the complete destruction of Satan and

35. Ruether, *Gaia & God*, 77.
36. Ruether, *Gaia & God*, 77. One wonders what Mary Cary's reaction was when Charles II was enthroned, the English monarchy restored and the established church returned to its position of power.
37. The date was later recalculated to October 1844, but, needless to say, his followers were disappointed.
38. Ruether, *Gaia & God*, 79.

the fiery purgation of the earth, which will then be recreated on a new and immortal basis.

Ruether's discussion of these two apocalypticists provides a brief but fascinating insight into the apocalyptic movement in Puritan England and North America in particular. She succeeds in illustrating that apocalyptic thought in the early church quickly became the domain of marginalised Christians, but has more recently gained popularity once again, albeit in more extreme forms of Christianity, as embodied by, for example, Mary Cary and Ellen White and the inheritors of their traditions. One of the strengths in Ruether's work is in drawing the attention of her readers to the dangers of radical apocalypticism. Its influence on American society and politics is not to be underestimated, as Ruether points out. The cold war of the 1950s to the 1980s and the ever present threat of nuclear war have provided Cary and White's theological successors with ample material for apocalyptic speculation. Likewise the founding of the state of Israel in 1948, and the expansion to the Gaza Strip and the West Bank in 1967, have provoked a fresh round of premillennial, apocalyptic predictions. Cary's work is of interest here, because of her vision of the Jews converting to Christianity and returning to their home land of Palestine. This restoration of a Jewish state and the rebuilding of the Temple has been a feature of Protestant apocalyptic thought since Cary's time. Even the failure of the Jews to convert to Christianity has not caused the vision to be abandoned. Rather the Jews will remain in an unbelieving state, but when the process of returning to Palestine is completed, one hundred and forty-four thousand Jews will be converted to 'true' Christianity and all false believers, that is Jews, Muslims, pagans, Communists and 'false' Christians, will be wiped out in the war of Armageddon between God and Satan. Meanwhile 'the true believers [will] be raptured up to heaven and thus kept safe during this war'.[39] Ruether also alerts the reader to the fact that even more recent events, such as the 1990 Persian Gulf crisis, have also been interpreted under this apocalyptic paradigm. Since the publication of Ruether's *Gaia & God* in 1993, a number of other events have also been interpreted in this way. A brief internet search suggests the more recent 2001 attack on the *Twin Towers* in New York and the subsequent wars in Afghanistan and Iraq have also been incorporated into various apocalyptic speculations.

39. Ruether, *Gaia & God*, 81.

Ruether's work is helpful in the way it highlights the dangers of apocalyptic thinking. Interestingly she does not explore, at least not in any detail, the development of the idea of 'the rapture'.[40] This is directly relevant to any discussion of the ecological impact of apocalypticism, because it promotes the concept of some Christians being 'raptured', or snatched, up to God before the general resurrection. The individual Christian leaves behind the world and all its concerns, including ecological concerns, and is taken to be with God. The connection between the individual and the world is suddenly and irreparably broken. The key publication promoting this concept has been the *Scofield Reference Bible* (published 1902–1909, revised 1917, 1966). This publication promotes the system of biblical interpretation known as *dispensationalism*, which has its roots in the preaching of John Nelson Darby of Dublin (1800–1882), one of the early leaders of the Plymouth Brethren; it has been highly influential in America. In the pre-millennial dispensationalist interpretation of the rapture, some people, *who are still alive*, will be caught up into heaven. It is a rather odd understanding of eschatology, which *almost* seems to be some kind of theological mutation of the doctrine of the Assumption of the Blessed Virgin Mary or the ascension of Elijah. The context, though, is quite different from those two examples, because in the rapture individuals are simply snatched away. There is no sense of the rapture being the final fulfilment of a long, godly life, as can be argued in the case of Elijah and Mary. Rather the believing Christian individual may be taken at any time. It is one example of a non-biblical, non-traditional doctrine being accepted as a part of fundamentalist Protestant theology in an uncritical way.[41]

Despite the value of Ruether's criticism of apocalypticism and its dangers, her discussion of the issue remains one sided. Her work does nothing to help the reader see the potential helpfulness of apocalypticism in providing a model for ecological eschatology. Ruether dismisses even the possibility of apocalyptic thought being of any use in environmental theology. But, again, it is the question, or charge, of oversimplification which Ruether must answer. The problem with Ruether's writing on this topic is that her examination of historical fundamentalist apocalypticism then

40. See John O'Neill, 'Dispensationalism' and Geoffrey Wainwright, 'Rapture' in *A New Dictionary of Christian Theology*, edited by A Richardson and J Bowden (London: SCM, 1983), 158 & 485.

41. To be fair, the concept of the 'rapture' seems to have come from a particularly narrow interpretation of 1 Thessalonians 4:17 and especially Matthew 24: 40 & 41 and Luke 17: 34 & 35.

limits any appreciation of the strengths of apocalyptic thought. We see no discussion of contemporary apocalyptic writing and are left only to ponder a very fundamentalist approach. The implication is that all apocalypticists are the theological descendents of Mary Cary and Ellen White.[42]

This is unfortunate, because apocalyptic thought is far more varied, subtle and nuanced than Ruether acknowledges, as both Ulrich Körtner and Jürgen Moltmann[43] have demonstrated. An examination of their work will form the basis of my next two chapters. It will suffice to say here that there is nothing necessarily inevitable about the rejection *per se* of the world arising from apocalyptic thought.

2.5 False dichotomies

The question is whether Ruether has posed a false dichotomy between eschatology in general, and apocalyptic thought in particular, and world affirming theology. Steven Bouma-Prediger points out that not all of Ruether's writing poses the false dichotomy. Indeed in her liturgical guide, *Women-Church: Theology and Practice*, she suggests that 'belief in immortal life need *not* necessarily entail a failure to value the earth. A creation-affirming theology is fully compatible with belief in "imperishable life" or life after death'.[44] Her suggestion in *Women-Church: Theology and Practice* is the exception, however, and the position she affirms in *Gaia and God* is consistent with her other work. Her main expressed position is that any hope for some kind of ongoing existence takes away from the desire to take any world affirming action, particularly environmentally. This position has been expressed more recently in her essay 'Eco-feminism: The Challenge to Theology', in which she reiterates the thought that patriarchal Christianity has always desired to 'dominate' and 'separate'. By ruling over others, especially women and nature, men seek to secure their own invulnerability from want and death. This necessarily has profound implications for women and for nature, because it results in 'an endless process

42. Ruether, *Gaia & God*, 81–84.

43. See Ulrich Körtner, *The End of the World* (Westminster John Knox Press: Louisville, Kentucky, 1995) and Jürgen Moltmann, *The Coming of God* (Fortress Press: Minneapolis, 1996).

44. Steven Bouma-Prediger, *The Greening of Theology: The Ecological Models of Rosemary Radford Ruether, Joseph Sittler, and Jürgen Moltmann* (Atlanta: Scholars Press, 1995), 145. See Rosemary Radford Ruether, *Women-Church: Theology and Practice* (San Francisco: Harper and Row, 1986), 212.

of seeking to amass power at the expense of the rest of humans and the earth'.[45]

From Ruether's perspective the vision of ecological hope, to which she holds, stands in stark contrast to the system of patriarchal domination, in which men seek to escape their finitude. Her vision is that of:

> an ecological hope freed from false escapism and content to make common joys abundant and available to us all in the midst of those tragedies of limits, failures, and accidents that also should be equally shared, rather than heaped upon some in excess so a privileged few may imagine themselves immortal.[46]

The real salvation that is available to us in Ruether's paradigm is not some kind of future paradise in which tragedy and death are overcome, based on the false projection of an escape from vulnerability. Rather in rejecting both the 'myths of immortal and perfect beginnings and ends' which 'falsify our real possibilities' we come to an acceptance and longing for a community of 'mutual life-giving, where we can hold one another in the celebrative as well as the tragic moments of our common life as earth creatures'.[47]

Ruether's criticism of mainstream Christian eschatology depends on the premise that eschatology will inevitably distract human beings from proper concern for environmental action and social justice. It is this premise which Bouma-Prediger has quite rightly questioned.[48] Whilst there have been examples in the history of Christianity of other-worldly concerns distracting the faithful from environmental and social concerns, it is by no means true that this *has* to be the case. Bouma-Prediger is cor-

45. Rosemary Radford Ruether, 'Ecofeminism: The Challenge to Theology', in *Christianity and Ecology: Seeking the Well Being of Earth and Humans*, edited by D Hessel and R Radford Ruether (Cambridge, Massachusetts: Harvard University Press, 2000), 105.
46. Ruether, 'Ecofeminism: The Challenge to Theology', 106.
47. Ruether, 'Ecofeminism: The Challenge to Theology', 106.
48. Bouma-Prediger, *The Greening of Theology*, 143. See also Joan Martin, 'A Sacred Hope and Social Goal: Womanist Eschatology', in *Liberating Eschatology: Essays in Honor of Letty M Russell*, edited by M Farley and S Jones (Louisville, Kentucky: Westminister John Knox, 1999), 209–226. Martin gives an alternative feminist perspective, which views eschatological hope as a helpful source of inspiration for fighting oppression and social injustice. For a helpful summary of various ethical positions see Janet Parker and Roberta Richards, 'Christian Ethics and the Environmental Challenge', in *Theology for Earth Community*, edited by D Hessel (Maryknoll, NY: Orbis, 1996), 113–131.

rect in asserting that, 'Certainly another option is to formulate a *different*, more creation-affirming eschatology. Ruether seems to assume an unwarranted dualism between history and eschatology.'[49]

The vision of a peaceful world, rejoicing in abundant harvests, where there is no exploitation or oppression and where creation is cared for and delighted in for its own sake, is by no means necessarily contrary to a vision of a transformed cosmos, rejoicing in the unmediated presence of its Creator. It is precisely the desire to develop a more creation-affirming eschatology which inspired this book and which will form the basis of the remainder of the work.

2.6 'Ecofeminist Theocosmology' and the question of evil and finitude

Having rejected traditional Christian notions of some kind of continuing or even transformed existence after death, Ruether then develops her own type of 'theocosmology', which is presented as a replacement for eschatology and traditional theism. She wants to develop a theology, which seeks an imaginative solution to the paradox of the traditional opposites of God's nature: immanence versus transcendence, female versus male, relational and interactive versus dominating, pluriform and multicentred versus uniform and monocentred. Building on the new physics, in which matter is understood as 'energy moving in defined patterns of relationality',[50] she rightly sees the entire universe as being inter-connected and any observer being a part of the process of the universe itself. At the tiniest level, physicists now believe, we should no longer talk about particles. In some experiments physicists were discovering 'energy fields in which energy "events" appeared and disappeared. Particles appeared out of energy and dissolved back into energy.'[51] Moving from the tiniest particles, and the energy field which forms them, to the largest possible scale, she suggests that we can talk about 'the matrix of all interconnections of the whole universe.'[52] This matrix, suggests Ruether, operates with a certain amount of relationality.

There is then a fundamental connection between, as she describes the tiniest parts of the universe, 'the dancing void of energy that underlies the

49. Bouma-Prediger, *The Greening of Theology*, 143.
50. Ruether, *Gaia & God*, 248.
51. Ruether, *Gaia & God*, 248.
52. Ruether, *Gaia & God*, 248.

atomic structure of our bodies and the universe'[53] and the universe itself. We make sense of this connection and our place in the universe through reflective consciousness. But, she argues, human consciousness, as highly intense an experience of life as it is, is nevertheless not separable from the rest of creation. For too long, males, in particular, have seen themselves as somehow apart from the rest of creation. Human consciousness, then, needs to be understood as a highly intense experience of life. But it must not be understood as being *absolutely* different from other experiences within the natural world. The most we can assert is that it is *relatively* different. Human consciousness is 'sustained by a complex but fragile organism. Cut that organism at its vital centers, in the brain or in the heart, and the light of consciousness goes out, and with it our "self".'[54]

We come now to the heart of Ruether's eschatology, and also its fundamental limitation. Ruether quite rightly draws the attention of theologians to the crucial connection between human consciousness and the dance of energy at the tiniest atomic level, which organises itself in increasing complexity until it indeed forms the ability to reflect back on itself with self-awareness. This fundamental connection, Ruether asserts, means that any concept of survival of human consciousness apart from the dance of energy must be rejected as untenable and also destructive towards the earth and other human beings.

As I have argued, and will argue in later chapters, the connection between hope for life after death and the destruction of the earth, in other words a negative ecological ethic, is *not* inevitable.[55] But the greatest limitation of Ruether's work is not, in fact, the denial of some quite positive models of ecological ethics in the form of eschatology, but rather the limitation she places on the understanding of God. At the climax of her work on eschatology she simply asserts that we will disintegrate at death. We need to accept the transience of ourselves; we are merely a brief episode in a much larger cosmic drama. At the end of our lives we will simply be reabsorbed into the larger matrix of energy, providing compost for the next generation: 'The cutting of the life center also means that our bodies disintegrate into organic matter, to enter the cycle of decomposition and

53. Ruether, *Gaia & God*, 249.
54. Ruether, *Gaia & God*, 251.
55. For a more balanced perspective on environmental ethics see Stephen Bouma-Prediger, *For the Beauty of the Earth: A Christian Vision for Creation Care* (Grand Rapids: Baker Academic, 2001).

recomposition as other entities'.[56] Thus the material composition of our bodies lives on in plants and animals, just as we ourselves are composed of substances that were once part of plants and animals stretching back through time to the very first life forms. Thus we are intimately linked, at a fundamental level, with the great Matrix of life, the living Gaia, and enjoy kinship with everything that has gone before us on earth and indeed even with the dust of exploding stars from long ago.

There is no doubt that Ruether's approach encourages an appreciation of the wider environment around us. Because of the intimate connection with all living things, the illusion of otherness, as Ruether puts it, is broken. We are connected with everything else and therefore to 'the ongoing creative Matrix of the whole'.[57] Aware of this connection, we value the personal of each created thing, from the small bird in the garden, to the ginger cat leaping with delight, to a tiny flower pushing through the soil to greet the same sun as Ruether.

The flaw in Ruether's argument, I would suggest, is that it is our realisation of our impermanence which allows us to see this connection between all living things and the 'Matrix of the whole'. She argues that it is the acceptance of the dissolution of our bodies into primal energy, becoming matter for new organisms, which then allows us to appreciate our connectedness, 'confident that our creative work will be nourishing to the community of life, even as we relinquish our small self back into the great Self'.[58] But the connection between the realisation of our impermanence and an appreciation of the value of every living thing is no more automatic than the connection between hope for life after death and a disdain for ecological concerns. Ruether assumes consequences of certain ways of thinking which simply cannot be assumed as being *necessarily* true. The realisation of our impermanence might just as easily lead to a kind of destructive nihilism. If there is no judgement, for example, no eternal consequence of my individual actions, then why not live for the moment and exploit everyone and everything to my advantage? The belief that this life is all we have can elicit *any number* of personal responses and not all of them, by any means, are ecologically positive.

There is, however, a more fundamental problem with Ruether's 'ecofeminist theocosmology'. Ruether's model of a creation affirming theology, in which 'Gaia' restores the ecological balance, presents an overwhelm-

56. Ruether, *Gaia & God*, 252.
57. Ruether, *Gaia & God*, 253.
58. Ruether, *Gaia & God*, 253.

ingly positive view of creation. This is not necessarily a sustainable point of view, theologically. An examination of Ernst Conradie's work on this subject will present an alternative perspective in a later chapter.[59] Moltmann, too, questions whether this model is helpful. Indeed he specifically addresses and criticises Ruether's ideas, as I shall discuss in chapter four. Even if Ruether's suggested program of environmental and political action were to be successful, a program she outlines in the final chapter of *Gaia & God*, 'Creating a Healed World: Spirituality and Politics', there is still the question of the finitude of creation itself. Whilst Ruether's program of reforming society and the attitude of human beings to the earth is inspirational in its own way, it avoids the difficult questions posed by social justice, evolutionary biology and the second law of thermodynamics. With respect to social justice, what do we say about those people whose lives and opportunities were brutally cut short? Are they simply forgotten, having been reabsorbed into the great Matrix of life? Is their contribution in fact simply a brief stay on earth, before providing compost for the next generation of life? With respect to evolution, what can we say about the dinosaurs and countless other forms of life which have since disappeared? Having had their turn, are they also forgotten, or perhaps simply remembered (in the absolute basic understanding of memory) as having made room for new forms of life? Finally, and perhaps most crucially, with respect to the second law of thermodynamics, what real hope is there for the earth and the dynamic Matrix of life, Gaia, if the earth and the cosmos itself have a finite life? All of these crucial questions remain unanswered by Ruether's model of eschatology. Her failure to address them adequately means her very model is flawed at a basic level.

It is interesting that Ruether makes a distinction in her work between domination and distortion, on the one hand, and tragedy and death on the other. She views tragedy and death as both natural and inevitable and argues that we 'must give up the future paradise where tragedy and death are overcome'.[60] The real salvation, then, she suggests, is not one free of tragedy and death, but rather one in which we live in a community of mutual life-giving, in which we hold one another in the wonderful moments of celebration as well as the tragic moments of despair. Thus the 'tragedies of limits, failures, and accidents' are equally shared, 'rather than

59. Ernst Conradie, *Hope for the earth: Vistas on a new century* (University of the Western Cape: Bellville, 2000).
60. Ruether, 'Ecofeminism: The Challenge to Theology', 106.

heaped upon some in excess so a privileged few may imagine themselves immortal'.[61]

The obvious limitation of this eschatology is that it is confined to a very limited view of soteriology. In other words, Ruether's eschatology is collapsed into soteriology and a fairly limited soteriology at that. In her model of eschatology there is no ultimate justice for the oppressed. Indeed there is no justice for those parts of the planet which are oppressed. Part of the focus of my work is working out an ecological eschatology, which means not just an eschatology which promotes an ecological ethic, but also one in which the whole creation shares in the coming transformation. Just as those whose lives have been cut short will be resurrected on the last day and experience justice, so too those parts of creation, which have been destroyed, will be remembered by God. The understanding of memory in this context is that sense of being fully recalled into God's eternal presence: the resurrection of all creation. The limitation of Ruether's eschatology is that there can be no justice for those parts of creation which have been destroyed. Since everything is destroyed—a consequence of our experience of finitude in the world—everything is lost.

2.7 Theology and ontology

The above concerns with Ruether's work all arise from the limitations of her doctrine of God. Theologically Ruether makes some questionable assertions. Her first assumption is that if creation is not somehow divine, in the sense of being a part of the divine life, that Christians will therefore feel free to exploit it. Despite her positive analysis of Irenaeus, she asserts that from the end of the third century, Christianity ceased to be concerned with the Hebraic vision of a transformed earth, as incorporated into Irenaeus' theology, and concerned itself with Christ establishing his power and reign through the political power of the church and Christian rulers.[62] I have already questioned some of her assumptions about the Christian doctrine of creation and whether Ruether here is making too simplistic a link between the rising influence and power of the church and a perceived lack of Christian environmentalism. Likewise Ruether automatically assumes a connection between the ontological separation of the divine God

61. Ruether, 'Ecofeminism: The Challenge to Theology', 106.
62. Ruether, *Gaia & God*, 237.

from the contingent creation and the exploitation of that creation, which is not necessarily automatic.

Be that as it may, her solution to the questions of theology, eschatology and ecology comes down too strongly on the side of God's immanence, despite everything that she says about maintaining a balance between God's transcendence and immanence. Ruether does in fact have a cosmic vision. She suggests that 'God in ecofeminist spirituality is the immanent source of life and the renewal of life that sustains the whole planetary and cosmic community'.[63] In this model the doctrine of the Trinity is then understood as 'the sustaining matrix of immanent relationality',[64] who sustains all life, both cosmic and planetary as well as social and personal.

But Ruether's doctrine of God is, in the end, one which limits God's transcendence, because there is no possibility of God doing something genuinely new. Though she uses the language of transcendence, in fact her understanding of God does not allow for that transcendence. In the end she fails to do what she sets out to do, which is to find an imaginative solution to the paradox of the traditional opposites of God's nature, particularly God's transcendence and immanence. In the end her theocosmology simply reverts to a restricted understanding of God, which in turn limits what God is perceived as being able to do. Indeed Ruether's own introductory comments set the scene for the limitations of her theology, when she writes that the reshaping of nature 'is finally governed by the finite limits of the interdependence of all life in that living system that is Gaia'.[65] The limitation of her theology, particularly her doctrine of God, is built on her unnecessary development of the concept of 'Gaia'. It is unclear why she needs to use the language of Gaia. Traditional trinitarian language for God, developed in an ecofeminist context by Elizabeth Johnson, for example, is more than adequate for exploring the very themes Ruether wants to explore. The language of Gaia is both unnecessary and, in the end, very restrictive.

2.8 Conclusion

Theology owes a great debt of gratitude to both Keller and Ruether for their feminist critique of theology, including eschatology. Their work on the potentially ecologically damaging consequences of apocalypticism

63. Ruether, 'Ecofeminism: The Challenge to Theology', 106.
64. Ruether, 'Ecofeminism: The Challenge to Theology', 107.
65. Ruether, *Gaia & God*, 31.

has resulted in eschatological theologians being forced to address the issue. Their criticism of mainstream Christian eschatology, that eschatology can distract human beings from proper concern for environmental action and social justice, is a fair one, even if I disagree with their conclusions. Likewise they have highlighted the connections between the oppression of women, and all 'lesser' beings, by patriarchal men, and the oppression of the environment. Of equal value is Keller's perceptive insight that capitalism and environmental destruction are closely linked. It is important too to comment on their respective visions for the future of the earth, in which we acknowledge our kinship with all of creation and our common life as earth creatures. The realisation and acceptance of the fact that we are intimately linked, at a fundamental level, with all life, is a profoundly impelling reason to struggle for ecological issues.

I have also made a number of criticisms of their work. The strongest criticism that can be made of both Keller and Ruether is not eschatological, as it happens. It is rather theological. Both Keller and Ruether are at such great pains to reject patriarchal, conservative and limited understandings of God, that they are left with models of God which are too focussed on God's immanence, despite Ruether's claims, in particular, that she wants to present a balanced understanding of God. The models of eschatology which emerge from Ruether and Keller's understanding of God are limited as a result of their doctrines of God. Neither allows for a sense of ultimate hope, because each fails to address the question of the finitude of the earth. Given that both the earth and the cosmos, at least in the current accepted scientific models, have a history which, in one way or another, will come to an end, both Keller and Ruether's models of eschatology are profoundly limited. The possible end of the world, whether through the inevitable entropy of a closed system, or through human beings' foolish actions in causing a nuclear or ecological disaster, seems not to be a question their eschatological models can cope with. Should the end of the world, or for that matter the end of the universe, come about, Keller and Ruether have nothing left to hope for. Neither theologian adequately deals with the question of justice, both for human beings, whose lives have been cut short, as well as those parts of creation devastated by ecological disaster.

An adequate ecological eschatology must address, with some hope, the questions of earthly and cosmic finitude. The problem of the end of the world is a real one, even though it may be a distant one, we sincerely hope. Apocalypticism is a real issue in the wider context of eschatology and so

I now turn in the next two chapters to the apocalyptic models proposed by Ulrich Körtner and Jürgen Moltmann, to examine whether they are ecological and thus provide a model for ecological eschatology.

Section II

Apocalypticism

Chapter Three
Ulrich Körtner

3.1 Introduction: a positive case for apocalypticism?

Is there a positive case for Christian apocalypticism, given both Keller and Ruether's assertions about the link between apocalypticism and destructive environmental activities? This section will examine whether there are positive, environmentally friendly themes within Christian apocalypticism. By examining the work of Jürgen Moltmann, one of theology's pre-eminent writers on apocalyptic eschatology, and Ulrich Körtner's key dissertation on the topic, I hope to demonstrate that apocalyptic thought has the potential to guide us towards a genuinely constructive model of ecological eschatology. This section will also demonstrate that apocalyptic thought is far more subtle and nuanced than Keller and Ruether allow for in their dismissal of it. Whilst certain models of apocalyptic theology can indeed be highly dangerous, there is nevertheless great variety in apocalyptic writing, just as there is in any other theological movement.

3.2 Liberating anxiety

Perhaps the undervalued quality of apocalypticism is its capacity to sharpen our focus, to snap us out of our blindness to the danger around us. Both Keller and Ruether build their criticism of apocalyptic thought on the understanding that the hope for escape from their world lessens our concern for it. It is like a drug which enables us to escape into the realm of fantasy when faced with the reality of potential environmental destruction. Körtner does not share their negative evaluation of apocalypticism. He asserts that we live in an age of anxiety. Yet, precisely because of the incapacity of people even to think about the potential for destruction for more than a few moments, we close our eyes to the approaching doom and try not to think about it! We are caught in the 'trance of the dance of death',[1] as Körtner describes it. In fact, he maintains, to be precise, we are not living in an age of anxiety, but an age *incapable* of anxiety. As a society we have repressed the fear of the end, both on a personal level in terms of the way we deal with death, and also on a communal and global level in the way we deal with (or fail to deal with) the possibility of the end of life as we know it on the planet.

Körtner makes a convincing argument in asserting that we are suffering from apocalyptic blindness, which is the result of repressed fear of the end. The possible positive impact of apocalyptic anxiety, what he calls liberating anxiety, has been suppressed by 'contemporary belief in progress, an understanding of progress and growth recognising no limits. It does not believe in an end, nor does it perceive any end. The notion of an end is itself inconceivable. Even the idea of one's own death is suppressed.'[2] The veracity of Körtner's hypothesis is easily tested by a quick examination of any capitalist society. This idea of progress, or growth, is indeed all pervasive in Western, capitalist societies. In Australia, for example, we measure the very health of our economy on the basis of *continual* growth. The idea that it all might come to an end, or *even* that continual growth might not be perpetually sustainable, seems to have been completely suppressed. There are always new markets to be found, new economies to be explored (or, perhaps more accurately, exploited). The excitement engendered by the opening up of China's economy, for example, can be explained by the perception that China's is an *almost limitless* market. Directors of Western companies can rest secure in the knowledge that, should they be successful in breaking into China's marketplace, they will have well and truly retired before China's potential is exhausted.

Apocalyptic anxiety can shake us out of this illusion. The approaching doom must be faced with open eyes and apocalyptic thought enables us to do this. It both opens our eyes to the reality of the end, as well as gives us the capacity to keep them open. Thus apocalyptic thought, at its best, helps us both cast off the false promise of endless progress, as well as face up to the end without feeling overwhelmed by it. To do so, however, theology must face up to its apocalyptic heritage and 'clarify its relationship to that heritage'.[3] Otherwise theology will either drift into some of the traps of apocalyptic thought itself (which I discuss below) or fall into a sense

1. Ulrich Körtner, *The End of the World: A Theological Interpretation* (Louisville, Kentucky: Westminster John Knox Press, 1995), 2.
2. Körtner, *The End of the World*, 3.
3. Körtner, *The End of the World*, 7.

of nihilistic hopelessness, or else simply revert to a state of apocalyptic blindness.

The danger of falling into a state of nihilism is real enough. Having been awakened from apocalyptic blindness, the cruel reality of the possibility of the end can be overwhelming. Körtner makes interesting use in this respect of the image of the sinking *Titanic*, which, he suggests, 'etched itself deeply into the consciousness of our century'.[4] The image of the sinking *Titanic* became a metaphor for existence under the continuing threat of nuclear destruction. What the twentieth century viewers of this particular image of the *Titanic* came to realise was that they were more than merely spectators. The threatened end of the world permits no spectators, not even theologians, who are included among those who are threatened.

3.3 The ambiguity of apocalypticism

Nevertheless we do need to adopt an air of caution when considering whether apocalyptic theology has anything positive to contribute to ecological eschatology. Whilst Ruether in particular does not allow for the positive elements of apocalyptic thought, Ulrich Körtner has pointed out that apocalyptic thought is best described as ambiguous. In other words we can speak of both positive and negative apocalyptic thought, with considerable variation within those two positions themselves.

The danger of certain forms of apocalypticism should be all too apparent to people who have lived with the perceived increasing threat of terrorism. Whatever the motivation of terrorism, certainly a common thread seems to be the desire to plunge the world into chaos and, at the same time, secure one's own eternal destiny. This book will not examine the particular fundamentalist interpretations of the *Koran* which claim to support suicide bombers and the like. However, the apocalyptic themes are common to parts of Christianity, Islam and Judaism. Körtner asserts that the danger of some forms of apocalypticism is that any individual, having been awakened from apocalyptic blindness, might in that very instance begin to long for and even bring about the kind of destruction threatened. Apocalyptic thinking needs careful examination, because who knows whether the apocalyptic anxiety engendered by apocalyptic

4. Körtner, *The End of the World*, 6.

thought might not, in its cynical frenzy, give 'what is already falling into the abyss a shove and then want[s] to leap into the abyss itself?'[5]

A careful examination of the terms is needed at this point. Whilst Körtner maintains that we can speak of both positive and negative apocalypticism, neither term should automatically be associated with concepts of good or bad. This underlines my point about the importance of recognising the subtlety of apocalyptic thought and the distinctions within it. Körtner defines negative apocalypticism as that movement which is capable only of verbalising apocalyptic world anxiety, but, unlike positive apocalypticism, is unable to nurture any hope in the face of the end. Yet negative apocalypticism can, somewhat ironically, be quite hopeful, in that it consists in *preventing* the end, provided such apocalyptic thought 'does not simply resign itself to anticipating the impending end'.[6] If the end is understood as the destruction of the world, then preventing the end might be seen as a good objective.

Negative apocalypticism can also play an important critical role within apocalypticism as a whole. Körtner uses the example of the potential for nuclear destruction as a means to illustrate the important and good impact negative apocalypticism can have in apocalyptic thinking. Precisely through its consideration of the possibility of nuclear destruction, and its lack of hope concerning the end of the world brought about by such destruction, negative apocalypticism *can* promote a socially critical movement, which actively strives against the possibility of nuclear destruction. Conversely positive apocalypticism can become rigorously positive in the form of fundamentalism and indeed put its hopes in nuclear doom, which is a criticism Körtner makes of the American evangelist Jerry Falwell. Other writers including Rosemary Radford Ruether have also highlighted the connection between this kind of fundamentalism and American conservative politics. Elements within Ronald Reagan's administration seemed to have been influenced by such apocalyptic thinking.[7] The connection

5. Körtner, *The End of the World*, 4.
6. Körtner, *The End of the World*, 198.
7. Ruether, *Gaia & God*, 81. For a fascinating insight into Reagan's personal opinions on the Apocalypse see also G Vidal, 'Armageddon', in *United States: Essays 1952–1992* (London: Abacus, 1993), 995–1006. Reagan believed, apparently with some fervour, that Armageddon was near. Moreover, Secretary of Interior, James Watt, and Secretary of Defence, Caspar Weinberger, 'expressed interests similar to Reagan about the imminence of the coming of Christ and the end of the world'. Daniel Wojcik, *The End of the World as we know it: Faith, Fatalism, and Apocalypse in America* (New York: NYU Press, 1997), 30.

between this rigorously positive apocalypticism and American foreign policy is a frightening one. The fear engendered by this connection was not allayed by the American President at the time of writing, George W. Bush.

Using Körtner's definition of negative apocalypticism, I would assert that Rosemary Radford Ruether and Catherine Keller can both be described as fitting within the category of negative apocalypticism.[8] Given their grave concerns with the apocalyptic movement as a whole, this is highly ironic to say the least! Yet the description holds. Both have articulated an apocalyptic world anxiety within the framework of environmental destruction. Both have used apocalyptic language to describe the end. Both have considered the possibility of ecological catastrophe in the context of a critique of apocalyptic thought (what Körtner would describe as apocalyptic-critical apocalyptic). Both see hope existing not in the end of all things but in the prevention of the end.

A helpful way forward in appreciating the value of apocalyptic thought may initially lie in the proper understanding of the term itself. In fact the term does not imply any sense of hoping for catastrophe, in the sense of longing for the total destruction of the world. Rather, as Körtner reminds us, *apokalypsis* simply means 'uncloaking, unveiling, disclosure, uncovering, or revelation'.[9] Thus when we talk about *apokalypsis* we are simply talking about the unveiling of what is now hidden. In the etymological sense of the word, then, apocalyptic thought is that which enlightens. Körtner suggests that the term unveiling may be the most helpful to describe what *apokalypsis* means, with specific reference to the unveiling of the future. Critically, though, this unveiling of the future, specifically the end of the world, affects one's interpretation of the present as a direct result. Indeed, if apocalyptic thought has a great deal of affinity with the concept of enlightenment, then it is the present which is enlightened in view of the impending end. Both history and the present are interpreted from the perspective of that end. It is this very unveiling, in that sense of enlightenment, which allows reality to be interpreted as being in decline. The future end of the world intrudes into the present time, because the hermeneutical key to the present is that expected end. Körtner suggests

8. Though Keller prefers the term 'counter-apocalypse'. See Catherine Keller, 'Postmodern "Nature," Feminism and Community', in *Theology for Earth Community*, edited by D Hessel (Maryland, NY: Orbis, 1996), 93–102.

9. Körtner, *The End of the World*, 199.

that 'the end of the world is not the end result, but rather the point of departure for the apocalyptic unveiling of reality'.[10]

The question then arises whether the unveiling of this future is *ultimately* a positive event for the world. Körtner convincingly argues that it can be. In this respect a key question will be whether Jürgen Moltmann's theology fits into the category of being an apocalyptic theology which is ecologically constructive. Körtner suggests that the message of apocalyptic thought is potentially liberating. When confronted with a reality which is closed off from all sides, apocalyptic thought, at least in its positive mode, provides liberation both from world anxiety and indeed from the world provoking that anxiety. Positive apocalyptic thought 'interprets the end as a transition and understands reality from the perspective of the end of the world interpreted thus'.[11] Thus apocalyptic thought negates the totally closed quality of reality. The future is not closed off and there is nothing inherently inescapable about it. For positive apocalyptic thought the end of the world is a kind of transition and is therefore the sublation of that sense of inescapability. Yet even negative apocalyptic thought can contribute something to the liberation from world anxiety in that respect. Though the world itself would be totally destroyed, the end of the world (from the perspective of negative apocalyptic thought) nevertheless would constitute liberation from the totally closed quality of the world. Even negative apocalyptic thought, therefore, views the end of the world as an egress, both spatially and in the sense of finality. 'Even the simple collapse of the world would break through the totally closed quality of the world, albeit at the price of its complete destruction. Even the end of the world as pure destruction constitutes the negation of inescapability'.[12] Thus, even though the destruction of the world would mean the complete loss of the future and the destruction of the ego, it would simultaneously mean the negation of a totally closed reality.

Körtner's use here of the language of 'inescapability' is revealing and unfortunate, because it contains more than a suggestion of 'flight from the world', which is precisely the main criticism levelled at apocalyptic thought by Keller and Ruether. What Körtner therefore needs to establish is whether apocalyptic thought, particularly positive apocalyptic thought, can indeed contribute to positive environmental action. In this respect his examination of damaged apocalyptic thought is helpful. So too is his un-

10. Körtner, *The End of the World*, 203.
11. Körtner, *The End of the World*, 204.
12. Körtner, *The End of the World*, 204.

derstanding of the 'sublation' of apocalyptic thought, which I discuss below. But even to use the language of 'inescapability' is fraught with danger in the context of apocalypticism. Perhaps the concept of a 'closed system', such as Jürgen Moltmann develops,[13] would be a better use of language. Language is powerful in any discussion and, given the eco-feminist criticism of eschatology and apocalypticism, which is that these movements within Christianity are all about fleeing from the earth rather than working ecologically for the earth, Körtner would be well advised to avoid the language of escaping.

In an earlier chapter Körtner does in fact address the issue of flight from the world, which makes his use of such language all the more confusing to the reader. However, his earlier work does address some of Ruether's concerns about the psychology of the end of the world. One of the key theses of his work is that 'apocalyptic is an attempt at overcoming total world anxiety through world negation'.[14] But his thesis depends on the rejection of the idea that the understanding of world negation as an attempt at flight is necessarily the *only* interpretation of world negation. In other words escapism is but one possible interpretation of apocalyptic thought. Certainly under the rubric of flight from the world there have been many movements, some of them apocalyptically driven. But, as Körtner points out, apocalyptic thought is not the only motivation for fleeing the world. Monastic asceticism comes from a quite different perspective from apocalyptic thought and yet has quite a specific focus on withdrawal from the world. However, whether Körtner is totally correct to associate monastic asceticism with world escapism is a debatable point. Often, indeed usually, the very cloistered nature of certain religious orders allows them to focus on the world in a very deliberate way, namely, through prayer *for* the world. Far from escaping the world these religious communities remain very much aware of world events, so that they can offer deliberate prayers for specific events and situations.

Nevertheless Körtner is correct in asserting that there are subtle variations under the category of world negation, which he views as the very essence of apocalyptic thought. Only one variation can be defined as escapism. World negation can in fact lead not to escapism, but world assault. This can be understood as either world affirmation or world rejection. Thus, depending on the perspective assumed, the world can either

---

13. See Jürgen Moltmann, *God in Creation: An Ecological Doctrine of Creation* (London: SCM, 1985), 212–214.
14. Körtner, *The End of the World*, 99.

be formed into something better, or totally rejected and abandoned to the wrath of God. World negation can, therefore, have very positive consequences for the world, because it allows new possibilities to break into the world. If apocalyptic thought is essentially world negation in the form of world denial, as Körtner asserts, then one possibility is that it can assault the world in a very positive way. It refuses to accept that the current world must inevitably assume the form that it currently does. It refuses to accept that the world must forever assume a dead end structure, and allows for the in-breaking of something radically new. Far from being a form of escapism, in this instance apocalyptic thought, as a form of world denial, longs for the ultimate transformation of the world, albeit beginning with the destruction of the world.

Having made the point about the careful use of language and terms, I must agree with Körtner when he asserts that apocalyptic thought can be highly political as well. The possibility within apocalyptic thought for political comment has both positive and negative potential (like apocalypticism as a whole). It can just as easily lead to political obscurantism as it can to functioning as a prophetic mouthpiece for the powerless and oppressed. The inherent danger is that apocalypticism will be ostracised by the world or isolated of its own accord and become ideological, meaning in this context incapable of self-critique. Apocalypticism is always in danger of ideological hardening (like any movement). It is therefore in danger of becoming what Körtner terms 'damaged apocalyptic'.[15] But, if apocalypticism avoids this and remains open to the necessary self-criticism, then it is capable of being a powerful political tool. The possibility of the catastrophic end of the world can cast a powerful light on the rest of reality, exposing all of reality, particularly the structures of power, to light. Körtner uses a helpful metaphor from photography here to suggest that the structures of power are over-exposed so that their complexity is uncovered and revealed and 'reduced to a dualism of black and white, good and evil, light and darkness'.[16] This simplification which results from the powerful light of the possible catastrophe is highly illuminating; the world unveiled by the apocalyptic vision can no longer dissemble the truth or remain in a perverted state, where evil is portrayed as good and vice versa.

Körtner has neatly illustrated one of the great contributions apocalypticism can make to the community. I would add that, just as anecdotal evidence suggests that, when individuals narrowly avoid death in an accident

15. Körtner, *The End of the World*, 14.
16. Körtner, *The End of the World*, 206.

or through illness, it can bring into much sharper focus the rest of their lives, so too the apocalyptic unveiling of the end of the world can have a similar effect on a larger scale.

To achieve this level of fruitful political contribution to the wider community around it, apocalypticism must remain open to two key dialogues. Firstly, it must be capable of the inner apocalyptic critique of apocalypticism, which emerges when the dialectic between positive and negative apocalypticism is explored. Secondly, it must also remain dependent upon critiques from external sources, which essentially means remaining in communication with 'non-apocalyptic experiences of existence and interpretations of the world'.[17]

Having established the ambiguity of apocalyptic thought, Körtner is then at pains to suggest that the way forward cannot involve playing one of its basic forms against the other and taking the side of either positive or negative apocalyptic. This is precisely the criticism I have levelled at Keller and Ruether: the tendency of both of them to adopt a completely anti-eschatological position, or what Körtner would describe as a negative apocalyptic position. But Körtner suggests that the way forward lies in the self-critique of apocalypticism, within the framework of the dialectic of positive and negative apocalypticism. This self-critique then promotes a critical discourse, leading eventually to what he terms the 'sublation'[18] of apocalyptic thought.

3.4 The repression of apocalypticism

Before we come to an examination of Körtner's proposed way forward, it is important to acknowledge that apocalyptic thought has, certainly for much of the history of Christianity, been marginalised.[19] This was Catherine Keller's point as well, as discussed in chapter one. This marginalisation has contributed to apocalyptic thought being dominated to a certain extent by fundamentalist thinking, leading to a total rejection of the world. Apocalyptic thinking was abandoned by other theological disciplines, resulting in peripheral church groups and sects outside the church administering 'this unwanted heir in their own fashion'.[20]

17. Körtner, *The End of the World*, 207.
18. Körtner, *The End of the World*, 256.
19. Though it has remained powerful despite, or perhaps because of, its marginalisation.
20. Körtner, *The End of the World*, 12.

The danger of our Christian apocalyptic heritage remaining the sole property of fringe groups of fundamentalists is threefold, it seems to me. Firstly, as Körtner shows, there are many strengths and benefits in the apocalyptic tradition, some of which are being discussed in this chapter. Secondly, as I shall discuss, apocalypticism has been a part of the Christian expression of reality since the origins of Christianity, a fact which militates against its total rejection by mainstream churches. Finally, and not insignificantly, there is a real danger of the extreme understanding of apocalyptic thought, as promoted by these fringe groups, making its way *back* into the mainstream churches. Without the benefit of contemporary systematic theological discussion, this extreme version of apocalypticism then becomes promoted as 'the Christian position', with the corresponding rejection of those who do not hold to it.

Körtner reiterates the point that the danger of fundamentalist apocalypticism is not a danger confined to a few strange historical movements or isolated thinkers. Whilst he does list some nineteenth century attempts at identifying particular historical events within an apocalyptic world view, Körtner also notes an extensive and increasing influence on recent and contemporary American politics. In this context the apocalyptic thought being promoted is anti-Communist and ultraconservative. Where the appropriation of apocalyptic thought is ill-considered the results can be catastrophic. Uncritical appropriation of apocalyptic thought often constitutes a *sacrificium intellectus*, he suggests, and represents apocalyptic thought 'that has misunderstood itself and fallen prey to the power of its own rhetoric'.[21]

The repression of apocalyptic thought provides the theologians with an interesting dilemma, in the sense that one must question whether apocalypticism is a central, or fundamental, part of the early Christian world view. The position which Körtner moves towards is that of the sublation of apocalyptic thought. However, he first examines the history of apocalypticism within a history of Christian thought. Körtner reminds us that some Christian thinkers identify apocalypticism as a defining characteristic of early Christianity, which was abandoned by later generations, resulting in the suppression of apocalyptic thought. Franz Overbeck is one such nineteenth century writer, who asserted that apocalyptic world negation was an essential part of early Christianity. The early Christian faith was faith in an imminent end of the world. Yet, according to Overbeck, this

21. Körtner, *The End of the World*, 11.

thoroughly apocalyptic Christianity was betrayed by a faith in progress and the future orientation of the present.[22]

A brief history of apocalyptic thought and its acceptance or otherwise by mainstream Christianity illustrates just how fraught the whole question of apocalypticism is. Even at the start of the Christian movement, apocalyptic thought was quickly marginalised. One sees this readily, Körtner asserts, with the dispute between Montanism and the established church, which was based around an emerging monarchical episcopate. The charismatic element of Montanism included a high regard for prophetic utterances, based around the expectation of the imminent end of the world and the 'descent of the heavenly Jerusalem onto a mountain near Pepuza and Thymion in Phrygia'.[23] Almost inevitably the group clashed with orthodox Christianity and this Phrygian prophecy was finally considered to be heretical. The marginalisation of apocalyptic thought became acute when Christianity emerged from its own marginal status as a small Jewish sect within the Roman Empire to become the official religion of the Empire under the Emperor Constantine. Körtner writes that the critical turning point for apocalypticism came with the rising influence of Eusebius of Caesarea, the court theologian under Constantine the Great. Rome began to be viewed as the power impeding any *premature* end of the world. Increasingly apocalyptic language began to be reversed and to stand in the service of worldly rule, rather than world negation.

Thus world negating apocalyptic thought was repressed within the church. It flourished in small groups in the Middle Ages and during the Reformation, both within the mainstream church and amongst smaller break away groups. But these groups were always in the minority and often persecuted. Although Joachim of Fiore's prophetic visions were apocalyptic and accepted within the Franciscan Order, later apocalyptic thinking during the period of the Reformation was often suppressed. The suppression of the Anabaptists, for example, also meant the suppression of apocalyptic thought, such as in the person of Thomas Müntzer. Thus even during the upheaval of the Reformation, when all of Catholic theology was scrutinised, apocalypticism did not achieve the acceptance that one might have expected, given the return to biblical theology that was encouraged. Even the place of apocalypticism within larger confessions of faith, for example Lutheranism, did not give it the required acceptance within theology as a whole.

22. See Körtner, *The End of the World*, 8–10.
23. Körtner, *The End of the World*, 13.

A monumental change occurred in the wide acceptance of Kant's philosophy and his description of 'religion within the limits of reason alone'. Apocalyptic traditions were, once again, suppressed within mainstream Christian thought, and effectively banished to smaller, more radical groups. In that sense apocalyptic thought has most often been connected with anti-establishment and anti-intellectual movements. The anti-Enlightenment theology of the eighteenth century and its successors is one example of that.[24] Apocalypticism has also been accepted more widely within the various revival movements, especially within Anglo-Saxon and German awakening movements. But even here the apocalyptic thought has been predominantly what Körtner terms damaged apocalypticism.

Körtner's key thesis is that the repression of apocalyptic thought by the mainstream church has been damaging, on the one hand, to apocalypticism itself, and, on the other, to the shape of mainstream theology and the church. His thesis is certainly convincing with respect to apocalypticism itself. The type of apocalyptic thought being espoused by some of the fringe sects, and even within certain more mainstream charismatic churches, is frightening to say the least. Whether the suppression of apocalyptic thought has been damaging to mainstream Christian theology as a whole is another question altogether. The key point of Körtner's thesis is to reassert the rightful place of apocalyptic thought within Christian theology and I will examine whether he does so convincingly.

Damaged apocalyptic thought is unable to confront the challenges placed before it by Kant's philosophy and that of his successors. Körtner writes that, 'As little as the optimism accompanying universal faith in progress is ready to confront its own anxiety, just as little is repressed, traumatised apocalyptic ready to confront a hostile world and the radical intellectual-historical changes that began with Kant.'[25] The failure to confront the challenges posed by Kant's philosophy, and the virtual abandonment of apocalyptic theology by the mainstream churches, left apocalypticism in the hands of the extremists, such as William Miller and the early Seventh Day Adventists.[26]

There is an interesting question for a theologian concerned with ecological eschatology: what were the social and environmental consequences of the influence of this damaged apocalyptic thought on the early Adventists. Körtner suggests that millions of people were affected by Miller's

24. See Körtner, *The End of the World*, 14.
25. Körtner, *The End of the World*, 14.
26. See Körtner, *The End of the World*, 14–15.

apocalyptic spirit and his predictions of the end of the world. 'Some of the faithful let produce rot in the fields, left their place of work, or closed their businesses, and many gave away all their possessions.'[27] Though Körtner does not do so, the environmental ramifications may be drawn from these events too. If the end of the world is nigh, with fiery judgment promised for unbelievers, with the purging of the world the means to achieve this, then any environmental action would be deemed a waste of time. What point would there be in worrying about the poisoning of our natural resources, or the elimination of native species, if the end of the world were upon us? In this context one can see very clearly the connection which Catherine Keller makes between apocalyptic thought and the absolute exploitation of natural resources by capitalist societies.

The example of the early Seventh Day Adventists is one example of damaged apocalypticism at work. It also illustrates the inadequate nature of Ruether's analysis of the apocalyptic movement. Whilst Körtner correctly identifies Miller and his successors (particularly EG White) as promoting a damaged version of apocalypticism, and therefore not being representative of all apocalyptic thought, Ruether makes no distinction at all between the fringe thoughts of Miller and apocalyptic thought as a whole. The subtle differences within apocalyptic thought are not addressed by her.

Thus Körtner's argument about the suppression of the apocalyptic tradition, and the resultant damage to apocalyptic thought in general, is persuasive. The critical question which emerges, however, for the purposes of this book, is whether the rehabilitation of the apocalyptic tradition will help in the development of an ecological eschatology. Both Ruether and Keller have argued that the very presence of the apocalyptic tradition within Christianity has resulted in mainstream Christian theology, particularly eschatology, being actively anti-ecology. It must be demonstrated that apocalyptic thought can say something positive about ecology, if it can have any place in ecological eschatology.

In what follows I will examine Körtner's work on the sublation of apocalyptic thought and Jürgen Moltmann's development of apocalyptic eschatology to discern whether apocalyptic thought can indeed play a positive part in the development of an ecological eschatology.

---

27. Körtner, *The End of the World*, 15.

3.5 The sublation of apocalyptic thought

In this context Körtner defines sublation as meaning that 'the Christian understanding of existence neither merely negates the apocalyptic understanding of existence nor shares it wholly without contradiction.'[28] In order to maintain this thesis, Körtner must firstly establish that Christianity, and specifically Christian theology, is necessarily apocalyptic. He acknowledges that un-apocalyptic Christianity in fact has a long tradition. I would add that this tradition has continued in various forms, not least in recent times in the eco-feminist writings of Ruether and Keller, and the work of Marcus Borg and the Jesus seminar.[29]

Körtner's work on the apocalyptic tradition within Christianity is interesting. One of the key points he makes is that it is the texts of the New Testament which provide us with the key to discerning the relationship between Christianity and apocalypticism. The question of the historical Jesus and his position over against apocalypticism is one which Körtner avoids. Quite apart from the constraints and focus of his own work, the question of the relation between the historical Jesus, the early church, the New Testament texts and the questions surrounding original sources and later redactional activity, would require an extraordinary methodological effort. Körtner simply highlights the issue, without engaging it. He highlights it principally because of the picture Schweitzer drew of the apocalyptic Jesus on the one hand, and, on the other, more recent hermeneutical attempts to rescue the New Testament Jesus from apocalypticism and draw a picture of him as un-apocalyptic.

Having established that his investigation will be confined to a brief examination of the New Testament texts, Körtner suggests in his summary that the influences of apocalyptic thinking in the New Testament are by no means confined to marginal writings or remote passages. Rather the entire New Testament stands under a strong apocalyptic influence and apocalyptic thought has influenced the most important New Testament writings. I do not propose to examine Körtner's hermeneutical method here, because it is his conclusions and their application which are of interest for this project. He is, however, most convincing when discussing one of his crucial conclusions, namely that the centre of the Christian faith is

28. Körtner, *The End of the World*, 256.

29. For an excellent summary of the search for the historical Jesus see Robert Crotty, *The Jesus Question: the historical search* (Blackburn Victoria: HarperCollins,1996), 135–179.

not in fact the expectation of the end of the world, but rather the salvific significance of the risen Christ. Certainly the significance of Jesus was interpreted in the early church using the language and imagery of apocalypticism. This applied particularly to statements concerning Jesus' resurrection. Moreover, suggests Körtner, the very structure of the early church was influenced by structural elements of apocalypticism, though, frustratingly, what these structural elements might be are not revealed by him.

Yet, whilst the language and imagery of apocalypticism was used in the early church, Körtner rejects Ernst Käsemann's thesis that Christianity represents some kind of unique version of apocalypticism, specifically, christologically modified apocalypticism. Central to his concluding work is Körtner's convincing argument that Christianity is not simply a variation of apocalypticism, as if it were some kind of sub-species—Christianity is considerably more than that. Moreover, the admittedly apocalyptic interpretation of the life, death and especially the resurrection of Jesus Christ (the apocalyptic interpretation of the kerygma of Jesus as the Christ) had, in turn, an impact on the interpretation of apocalypticism itself. The application of apocalyptic conceptual models in Christology is appropriate, suggests Körtner, but we must also acknowledge that the very use of such models then modifies apocalyptic thinking.[30]

The strength of Körtner's use of the term 'sublation' lies in the fact that he does not wish to present apocalypticism as some kind of solution to all of life's problems, including the previously identified problem of world anxiety. In this context anxiety, or, as I would suggest, doubt, is understood as a symptom of unbelief, while faith is confused with the feeling of being free from anxiety. Körtner perceptively identifies a tendency within Christianity to promote the concept of faith as the solution to everything, whilst even the possibility of doubt is something to be rejected at all costs. But he insists that 'an apologetic theology of anxiety in the sense described makes anxiety the presupposition of faith and equates the latter with freedom from all anxiety . . . This kind of theological "anxiety business," including all its fairly sublime forms, should be summarily rejected.'[31] Interestingly Körtner considers any theological anthropology, which dismisses the connection between the capacity for anxiety and our disposition as created beings, as highly problematical. This returns the reader to one of

30. Körtner, *The End of the World*, 240.
31. Körtner, *The End of the World*, 242.

Körtner's key points, that 'the capacity for anxiety is a basic anthropological constant'.[32]

Körtner's discussion of 1 John 4:18a is of interest here ('There is no fear in love, but perfect love casts out fear'), because it is not world anxiety which is being cast out, but rather fear of divine judgement. The freedom from anxiety referred to in this passage is not one's experience of the world, or relationship with the world, but rather one's relationship with God. Thus the challenge of faith and its demands must not be misunderstood as a call to an anxiety-free life.

This has become a pressing issue within Christianity and is not one confined to the question of the end of the world, though damaged apocalypticism often goes hand in hand with rigid fundamentalism, which leaves no room for doubt. This increasing rigidity within even mainstream churches depends on an understanding of faith which, for our purposes, is at odds with Körtner's understanding of the role of apocalypticism within Christianity. I will explore the ramifications for Körtner's model of apocalypticism below.

One of the keys to Körtner's development of his apocalyptic theology is then the notion that, far from faith being a call to abandon anxiety, rather the Christian takes on an attitude of apocalyptic world anxiety. This anxiety is part of the disposition of those who also hope for deliverance. But this deliverance is seen precisely in the end of the world, which is seen positively, because it overcomes the inescapability of finitude.

Körtner's thesis, however, as he is at pains to point out, is that Christianity is not simply a variation within apocalypticism. Thus whilst the Christian faith stands within the framework of apocalyptic world anxiety, it also believes that 'the structure of reality disclosed by that anxiety has been broken through in the fate of Jesus'.[33] Yet the Christian faith also leads the believer into a position of anxiety, because the overcoming of the world, asserted to have happened through the cross and resurrection, is not visible—it is only believed. Christian courage, therefore, is precisely courage to be in a state of apocalyptic anxiety and to use language influenced by apocalypticism.

What Körtner therefore asserts is that the category of anxiety itself shows that the Christian understanding of existence is not un-apocalyptic, because it stands within the framework of apocalyptic anxiety. Thus a thesis of an un-apocalyptic Christianity is unsustainable. If we push Körtner's

32. Körtner, *The End of the World*, 242.
33. Körtner, *The End of the World*, 254.

argument a bit further, any Christian paradigm which is not apocalyptic is therefore not Christian, at least not in the sense that Körtner develops it. Yet neither does the Christian understanding of existence follow in an unbroken fashion the apocalyptic understanding of existence, because apocalyptic world anxiety is 'both provoked and simultaneously altered through the fate of Jesus, which culminates in the cross and resurrection'.[34]

The sublation of apocalypticism within Christianity then forms the central argument of Körtner's concluding comments. What is of interest to my work, however, is whether the sublation of apocalypticism can provide theology with a model for ecological eschatology.

3.6 Ecological ramifications of apocalypticism

For the Christian apocalypticist, for whom the sublation of apocalypticism is the guiding principle, the beginning of the end of the world has already occurred in the cross and resurrection of Christ. But the danger for the Christian apocalypticist is to interpret this beginning as, in fact, the end of history. That is to say, Christ is understood as the end of history and as the one who provides liberation from history. Here history is understood as the historical world, which is disastrous, but from which the Christian is now liberated, that is, disentangled or disengaged. Körtner sees this as falling under the charge of religious escapism.[35] From my ecological point of view, viewing Christ as the end of history would promote a *laissez faire* attitude to the environment, because nothing on this earth would matter any more. Körtner asserts, correctly, that it is not the Christian position to view Christ as the end of history *per se*. Indeed the New Testament writers use apocalyptic language which refers not just to

34. Körtner, *The End of the World*, 255.

35. However, it needs to be noted that a theology which understands Christ as the end of history need not necessarily lead to religious escapism. The theology of Pannenberg is a good example of a theology which understands Christ as the end of history, but which doesn't lead to escapism. 'Jesus Christ, then, is the Word of God as the quintessence of the divine plan for creation and history and of its end-time but already proleptic revelation.' Wolfhart Pannenberg, *Systematic Theology Volume 1* (Grand Rapids, Michigan: William B Eerdmans, 1991), 257. See also Wolfhart Pannenberg, *Systematic Theology Volume 3* (Grand Rapids, Michigan and Edinburgh: William B Eerdmans and T&T Clark, 1998) especially 527–555 and Ted Peters, 'Editor's Introduction: Pannenberg on Theology and Natural Science', in Wolfhart Pannenberg, *Toward a Theology of Nature: Essays on Science and Faith,* edited by T Peters (Louisville, Kentucky: Westminster John Knox Press, 1993), 12.

the past, but also to the present and the future. There is an *expectation* of the end of the world, which the thesis of Christ as the end of history would seem to ignore or marginalise. Körtner considers that the understanding of Christ as the end of history separates Christianity from apocalypticism, which he is keen to avoid.

This then leads to the question of what the key similarities and differences between non-Christian apocalypticism and apocalyptic Christianity are. The apocalyptic understanding of the world sees the world as the nexus of disaster. History is indeed an enclosed continuum of disaster, because the world itself is 'an enclosed time-space nexus of disaster'.[36] Salvation can therefore *only* come from outside the world and history. Given that the present and the past are, according to the apocalyptic understanding of the world, utterly devoid of any salvific experience, salvation can only come from the future. This salvation comes through catastrophe. Hope is placed entirely in the end of the closed, dead-end structure of the world.

The Christian understanding of salvation shares many things in common with the apocalyptic picture of salvation. It too looks for future deliverance. If we accept Körtner's claim that un-apocalyptic Christianity is untenable as a thesis, then the Christian understanding of salvation is also based on the catastrophic nature of this salvation. A fundamental difference, however, is that the Christian apocalypticist understands the cross and resurrection of Christ as a salvific event. This is critical, because history then becomes not just a self-enclosed continuum of disaster, but also the locus of an experience of salvation. History itself is redeemed in the Christ event. 'In the midst of the catastrophic nature of reality there is an experience of a counterworld of love, righteousness, and life.'[37]

The difference is quite profound and has ecological consequences, though Körtner's focus is quite different and he does not explore the ecological ramifications. But if the Christ event can be understood as an in-breaking of salvation, as Körtner does, then this world no longer remains simply a place of sin and disaster. Whilst the historical in-breaking of salvation must not be misunderstood as salvation's total realisation, nevertheless this in-breaking changes the character of history. It alters the nature of reality, such that the creation itself is a place of salvation. Unlike the purely apocalyptic experience of the world, which is characterised by God's remoteness and consigns God's presence in the world to a future,

36. Körtner, *The End of the World*, 258.
37. Körtner, *The End of the World*, 259.

catastrophic event, Christian theology sees God as having already begun the process of salvation in the world. Thus God is characterised as the coming God, a position made famous by Moltmann, understood most clearly by contrasting the terms *futurum* and *adventus*, translated *Zukunft* in German. The Latin term means 'coming towards', as does the German translation. The Christ event is an in-breaking of the future, which alters the fabric of reality itself, because God is near. God is no longer remote or confined to a future catastrophe.

Where the question of Christianity and its relation to apocalyptic thought becomes problematic, however, is precisely in its relation to the world. Non-Christian apocalypticism sees nothing positive about the world at all. Christian apocalypticism, on the other hand, understands the Christ event as having changed the fabric of reality. Yet precisely this understanding has led some theologians to reject the apocalyptic notion of the destruction of the world inaugurating any kind of salvific reign of God. Using Körtner's ideas, I suggest that the cross of Christ is a paradoxical symbol of the Christian understanding of the world. It is both a symbol of world negation (a world heading for destruction as symbolised in the tragedy of the cross) and at the same time it is a symbol of world affirmation (Jesus' fate is the in-breaking of God into this world). But faith can push on to world affirmation only through world negation. 'Faith is world affirmation that must repeatedly pass anew through apocalyptic world negation.'[38]

It is precisely at this point that Körtner's apocalypticism stumbles on the question of ecology. Whilst he quite helpfully defines faith as 'courage for questionable existence',[39] suggesting a limit even to eschatological thinking in the face of the absolute end, he also then suggests that the total destruction of humankind would constitute the extinguishing of human beings' memory of Jesus and of God. That is, there would remain no subject into whose memory God might recall himself. This would be different from the resurrection of Jesus, who was called by God into living memory for both God's self and for human beings. Thus, with the total annihilation of humankind, God, humankind and, by extension (though Körtner makes no mention of creation) all of creation would be forgotten. In this context there is an inseparable link between being dead and being forgotten. With the extinction of humankind, the dead will be totally forgotten.

38. Körtner, *The End of the World*, 265.
39. Körtner, *The End of the World*, 265.

'The dead would then have been as if they never were.'[40] Memory would be forever annihilated along with humankind itself. 'The annihilation of humankind would be the equivalent of the annihilation of all memory.'[41]

Here Körtner's theology reveals itself as being too anthropocentric. Even his concession that, in the event of the total destruction of humankind, God would nonetheless be able to recall into his memory his own humankind, creation seems to have been forgotten. The model proposed in the event of total destruction seems to allow no place for a renewed creation, or even a *remembered* creation.

Körtner's final ideas do, however, provide us with a launching point for a truly ecological eschatology, when he develops the notion of Christian engagement with the world, both in action and also in suffering. Faith involves not just the passive endurance of apocalyptic disaster, but also active co-suffering with the victims of history. These are remembered before God and thus the memory of the dead is not extinguished. Prayer, in the form of lament, petition and intercession, brings the dead before God, and is itself a protest against the absurd. Yet faith itself borders on the absurd, because it exists in the shadow of the threatening end.

Körtner's inclusion of the category of absurd here provides a fascinating point of contention within the mainstream churches (leaving aside for the purposes of this particular discussion the extremist fringes of Christianity). As I indicated above, the increasing rigidity of much popular theology has become a pressing theological and ecclesiastical issue. No longer, it would seem, is a position of doubt or even compromise acceptable.[42] In this environment of growing rigidity, the question must then be asked of Körtner whether his concluding ideas are helpful. This is not an abstract question, because any ecological eschatology must have the potential to have an impact amongst the lives of individuals. Ultimately it is debatable whether Körtner's theology passes this particular test.

A more critical fault in Körtner's theology, however, is that he fails to provide a model of eschatology which has ultimate hope in the shadow of the threatening end. His reticence is partly to do with the limits of epis-

40. Körtner, *The End of the World*, 274.
41. Körtner, *The End of the World*, 274.
42. For example, as evidenced in an Australian Anglican context by the *majority* of the current hierarchy of the Anglican Diocese of Sydney. For a fascinating discussion of this issue see Muriel Porter, *The New Puritans: The Rise of Fundamentalism in the Anglican Church* (Melbourne: Melbourne University Press, 2006), especially chapter one, '"True Believers": Completing the Reformation', 9–32.

temology and language. He is critical of Paul Tillich, whose theology he suggests opens itself up to the charge of escapism. Tillich suggests that God overcomes all negativity 'by eternally remembering the positive in all things while excluding the negative'.[43] Körtner suggests that Tillich's theology exceeds the limit of what we can say about the doctrine of God. There is a limit imposed upon us by the end. The end of humankind would constitute the end of God *for us*. Tillich's theology, he argues, attempts to elude, in an escapist fashion, the threat of destruction. But even if God survived the destruction of all that exists, this would be a *deus absconditus*, rather than the *deus pro nobis*, whom we know through the person of Jesus Christ. This God did not annul all negativity in himself, but rather on the cross falls 'into the vortex of nothingness'.[44] However, in more traditional theological understandings the theology of the cross is balanced by a theology of the resurrection. In Körtner's schema we are simply left with faith in dialogue with the absurd, and no substantial hope.

3.7 Conclusion

Körtner's work is valuable as a detailed analysis of apocalyptic thought. His work reveals that Keller and Ruether's dismissal of apocalyptic thought, as being completely anti-ecological, is too simplistic a position. Körtner points out that apocalyptic thought is best described as ambiguous, with considerable variation within the two main categories of positive and negative apocalyptic thought. But it is not necessarily anti-ecological. For example, apocalyptic anxiety can shake us out of the illusion of endless progress and enable us to face up to the end without feeling overwhelmed by it. What Körtner categorises as negative apocalypticism can, ironically, have quite ecologically positive consequences, because, as long as it does not resign itself to anticipating the end, it can help *prevent* the end. Negative apocalypticism can potentially be quite positive towards the ecology of our world, because it attempts to prevent the approaching doom. Positive apocalypticism is also potentially liberating, because it can provide liberation both from world anxiety and indeed from the world provoking that anxiety. Positive apocalypticism can be viewed, not as a form of escapism as Keller and Ruether insist, but rather as a form of world denial, which longs for the ultimate transformation of the world, albeit beginning

43. Körtner, *The End of the World*, 272.
44. Körtner, *The End of the World*, 273.

with the destruction of the world. Deliverance is seen in the end of the world, which is viewed positively, because it overcomes the inescapability of finitude.

Körtner's work is also valuable in pointing out that the repression of apocalyptic thought by the mainstream church has been damaging both to apocalypticism itself and to the shape of mainstream theology and the church. His proposal, that Christianity has always been shaped by apocalyptic thought, and has in turn shaped apocalyptic thought, is persuasive. Christian courage is courage to be in apocalyptic world anxiety and to use language influenced by apocalypticism. The 'sublation' of apocalypticism is the term Körtner uses to describe Christianity's acceptance of apocalyptic thought, but acceptance which is shaped by the cross and resurrection of Christ. The Christian apocalypticist understands the cross and resurrection of Christ as a salvific event.

Körtner ultimately fails to establish that the 'sublation' of apocalyptic thought can indeed contribute to positive environmental action. He fails to provide a model of apocalypticism which has ultimate hope in the shadow of the threatening end, because he suggests that the total destruction of humankind would constitute the extinguishing of human beings' memory of Jesus and of God. There would remain no subject into whose memory God might recall himself.

However, the failure of Körtner to answer satisfactorily the criticisms of apocalyptic thought made by Keller and Ruether does not mean that we must accept their rejection of apocalypticism on ecological grounds. I turn now to Jürgen Moltmann's work on apocalypticism to examine whether his work can contribute something to an ecological eschatology.

Chapter Four
Jürgen Moltmann

4.1 Introduction

Jürgen Moltmann is one of the most prolific recent writers in the area of eschatology and particularly apocalyptic eschatology. The question must be asked of Moltmann whether he indeed balances the dialectic of positive and negative apocalyptic thought and moves towards its sublation. Moreover, if he does find that balance, does Moltmann also provide us with an eschatological model which is ecologically constructive? The ambiguity which Körtner highlights surrounding apocalyptic thought needs to be addressed, if any apocalyptic model of eschatology is to have anything positive to contribute to the ecological debate. Does Moltmann's apocalyptic eschatology unveil reality in a way which is then constructive in an ecological sense?

4.2 The coming God

Like Körtner, Moltmann understands eschatology principally using the framework of *adventus*, and understands God as 'the Coming God'.[1] The key principle from which Moltmann works is that the eschaton is neither the future of time, in the sense of *futurum*, nor is it a concept of timeless eternity. Rather God comes from the future, bringing conversion, that is, new life, with him. Moltmann understands this conversion as being more than a mere interruption of time, in the sense that a prophet might interrupt the ordinary flow of events. Rather 'the future made present creates new conditions for possibilities in history'.[2] God's coming towards us means that he brings eternal life and eternal time. The coming of God

1. Jürgen Moltmann, *The Coming of God: Christian Eschatology* (Minneapolis: Fortress Press, 1995), 22.
2. Moltmann, *The Coming of God*, 22.

means a fundamental change in the conditions of time. Moltmann understands the 'transcendental future of time' as surpassing the present and the past. Consequently, when God comes from this future, he will gather up all of time and transform it. This concept of future time is crucial to Moltmann's work. I shall examine Moltmann's concept of time in more detail below, but it needs to be noted here that it does create problems for the importance of history and therefore the processes of creation. If God's eternity transforms all of time, what importance does the history of creation have? If the history and processes of creation should turn out to be unimportant, what motivation then do we have to care for creation at all?

God's coming towards us is also understood as meaning that God comes to dwell with us, in the world which will be transformed by God's presence in it. Thus not only is time fundamentally transformed, but so too is space.

The category Moltmann uses to explain the transformation of time and space is that of *novum*—the new thing. Moltmann suggests that this category is one which emerges from Israel's theological history and belongs to the messianic faith. God is understood as the creator of a new future. Moltmann sees this theme emerging particularly in the prophetic tradition, in which the category of hope in the new future promised by God takes over from the category of remembrance. However, the past is not forgotten. Rather 'what God sent forth in the past becomes the prologue to the future'.[3] Thus, in the context of the Hebrew Scriptures, the Exodus and Jerusalem become the foundation of hope for the new Exodus and the new Jerusalem. Moving then to the language of the New Testament, Moltmann suggests that we can understand the category of *novum* by looking at the raised Christ. The risen Christ, Moltmann insists, 'does not issue from the history of the old'.[4] Rather something new comes into being, something which is astonishing, because it could never have been expected.

The problem Moltmann faces here is the question of continuity. If creation is both *creatio nova* as well as *creatio ex vetere*, what is really left of the old creation? Does the old creation have any use? Perhaps more significantly for those of us concerned about ecology, if the transformed creation is something genuinely new, does it really matter what we do with the old creation? If God can create something new, can open up rich pos-

3. Moltmann, *The Coming of God*, 28.
4. Moltmann, *The Coming of God*, 28.

sibilities from the transcendental future of time, could not he simply fix whatever mess we create here on earth?

Moltmann's response to this potential criticism is to suggest that what is eschatologically new creates its own continuity. The *creatio nova* is precisely the new creation of this one. However, whereas this creation is perishing from its sin and its injustice, the new creation will have no room for such things. Thus this creation will indeed form the basis of the new creation, but only in the sense that it will be gathered up and created anew.

It is God's eschatological future which redeems us from the power of history and time. One can see here the influence of Moltmann's early experience of the Second World War and his exposure to the shifts in theology following the end of the First World War. The concept of any understanding of progress within history is one which Moltmann firmly rejects. Indubitably one sees here a rejection of progress in the light of twentieth century German history. Certainly part of Moltmann's rejection of history is to do with the perception that 'the power of history is exercised by the mighty'.[5] He is most convincing when he suggests that it is the mighty who press forward in history, in order to consolidate their power, to prolong their present state and then to expand it. Their future is projected and planned for, because, he suggests, only those with power can possibly project and plan. Despite the lack of optimism for the future, the pressure for progress is unchanged, because of the compulsive nature of competition.

The other factor in Moltmann's rejection of progress has to do with his rejection of evolution as a process which might contain the seeds of salvation. I will discuss Moltmann's attitude to evolution below, but suffice it to say there are internal contradictions within Moltmann's work on the subject of evolution.

Placed against the pressure of history, according to Moltmann, is the hope of God's eschatological future. God's future wins power over the present and, in doing so, opens up new possibilities, which are absent in the future of the 'mighty', because their future is entirely planned and therefore devoid of surprises. But God's future opens up the present and allows us to experience eternal moments, which interrupt time and let us pause in the midst of progress, in order that we might recognise the deadliness of progress. Unchecked, progress leads inexorably to catastrophe, whether economic, ecological, nuclear or genetic. But the arrival of an 'eternal

5. Moltmann, *The Coming of God*, 45.

moment' (understood from a messianic perspective, not a mystical one)[6] from God's transcendental future allows alternative developments to be imagined. Moltmann terms this redemption of the future from the power of history the *kairos* of conversion.[7]

One sees here in Moltmann's work the category of apocalypticism, as Körtner defines it, because the future is no longer static or closed; it is not inescapable, but rather is full of possibilities. Moltmann's work has the characteristics of negative apocalypticism, because he sees all-too-clearly the approaching catastrophe. Whether he also shares some of the characteristics of positive apocalypticism, is an interesting question. Moltmann certainly looks to the end and sees hope in the end. But, as I shall discuss, whether Moltmann can be understood as a true apocalypticist, at least in the way that Körtner defines one, is a debatable point. The possibility of total catastrophe is not one which Moltmann is prepared to consider, which suggests his eschatology may not be the *apocalyptic* eschatology he insists it is.[8]

4.3 Concern for creation

Moltmann's concern for creation within his eschatological schema can be readily established.[9] Whether, though, his eschatology evokes within the reader the desire to act ecologically, is another question altogether. Nevertheless his intentions are clear from the start. He spends quite some time discussing personal eschatology, including fascinating questions such as the doctrine of the immortal soul, the question of sin and death, the doctrine of purgatory, resurrection at death, reincarnation and the place of mourning and consolation within the Christian life. This work will not

6. See Moltmann, *The Coming of God*, 45. It's important to note Moltmann's distinction here between a 'mystical interpretation of "the moment" that "leaves nothing more to be desired"; and [...] the messianic interpretation, which throws open new perspectives, and discloses everything that is to be desired.'
7. Moltmann, *The Coming of God*, 46.
8. See Moltmann, 'Is Apocalyptic Eschatology Necessary?' in *The Coming of God*, 226–235.
9. See for example, Jürgen Moltmann, *God in Creation: An Ecological Doctrine of Creation* (London: SCM, 1985), especially 181–184, Jürgen Moltmann, 'The New Earth: Eschatological Ecology', in *The Coming of* God, 277–279, and Jürgen Moltmann, *The Way of Jesus Christ: Christology in Messianic Dimensions* (London: SCM, 1990), especially 302–305.

examine those themes in detail, except where they are directly relevant to ecological eschatology.

Moltmann's discussion of the possibility of individual resurrection at death reveals his ecological concerns. He ultimately rejects the concept of resurrection at death, on the grounds that human beings are in bodily solidarity with this world. Were we to be raised from the dead, we would be raised from this unredeemed world. But then our solidarity with this world would be broken. For Moltmann this concept is unacceptable, because it sunders our experience of the world from our raised identity. The raising of the dead must therefore happen as a part of the creation of God's new earth, in which death will be no more. He suggests that universal eschatology cannot be reduced to personal salvation. Personal perfecting and the perfecting of the world itself must coincide. Thus Moltmann rejects the concept of individual resurrection at death on the grounds of our being connected at a fundamental level with the world around us.[10]

The other telling reason Moltmann rejects the concept of individual resurrection at death concerns the question of time. Here we begin to see some inconsistencies emerge in Moltmann's work, because in rejecting the doctrine of resurrection at death, he also appears to qualify his own concept of time. But he then promotes this understanding of time when talking about the final consummation of all things: the coming of God. Here Moltmann rejects the possibility that resurrection occurs at death, because this would mean that all human beings, whenever they might have died, would encounter God immediately after their death. Each individual would be led directly to the eternal 'Day of the Lord', the moment when the final consummation takes place. Moltmann rejects this and suggests instead that there is an intermediate time, between Christ's resurrection and the general resurrection of the dead. In this time the dead are simply with Christ. This time is best understood as Christ's time, which is 'the time of love, the accepting, the transfiguring, the rectifying love that leads to eternal life'.[11] Somewhat ironically, given Moltmann's Reformed background, this then leads him to acknowledge that this concept of Christ's time is the true element in the doctrine of purgatory.

But Moltmann has created problems for himself precisely in rejecting the concept of resurrection at death, because his own understanding of the time of the 'Day of the Lord' would seem to argue against any notion

10. Moltmann, *The Coming of God*, 104.
11. Moltmann, *The Coming of God*, 106.

of intermediate time. This and other problems with Moltmann's concept of time will be discussed further below.

4.4 Millenarianism

Moltmann is a very interesting contemporary theologian because he discusses some theological doctrines, which other ecologically minded theologians either dismiss or no longer consider important. His interest in millenarianism and apocalyptic eschatology fall into this category. As I have shown, both theological movements have been criticised and dismissed by Ruether as at best unhelpful and at worst highly destructive for the ecological movement. Yet Moltmann persists in exploring these categories. Whether he does so beneficially in an ecological sense, I shall now discuss.

Even before his seminal work on eschatology, *The Coming of God*, Moltmann had already begun to explore the place of millenarianism within eschatology. In an interesting article, in which Moltmann draws Thomas Aquinas and Joachim of Fiore into a continued conversation, Moltmann emphasises the balance to be found between the two positions. On the one hand there is Aquinas, whose vision of the perfection of all things lies indeed in the beatific vision and the transcendent nature of perfection. On the other Joachim's vision is far more historically based. He thinks historically, 'within the framework of his typological exegesis of history: promises call forth hopes, and these hopes call for the fulfilment of the promises'.[12] Joachim's work is not historical *per* se, but reflects his translation of biblical eschatology into a chronology of world history. However, there are problems with this approach, as Moltmann shows.

Moltmann is understandably wary of translating the biblical history of promise and its eschatology into a chronology of world history, but he is equally determined not to replace biblical eschatology with 'finalistic metaphysics'.[13] In order to achieve something of a balance between Joachim of Fiore and Thomas Aquinas, Moltmann explored the concept of *adventus*. Critically, for Moltmann all eschatology must be christological and Moltmann sees 'in the rule of the risen Christ the *historical* beginning of that eschatological kingdom in which God will be all in all'[14]

12. Jürgen Moltmann, *History and the Triune God: Contributions to Trinitarian Theology* (London: SCM, 1991), 94.
13. Moltmann, *History and the Triune God*, 95.
14. Moltmann, *History and the Triune God*, 95.

[my italics]. Already in the power of the Holy Spirit the parousia is present in a spiritual way, but the eschatological future will arrive in a series of particular presents or states. These ages, as Moltmann eventually calls them, overlap and are transitions in the one eschatological movement of history. The various ages are characterised and determined by particular signs: the historical age by the promise; the messianic age by the messiah; the spiritual age by the Spirit of the new creation; the chiliastic age by the kingdom of Christ; the eschatological age by the judgment and kingdom of God; and the eternal age by the feast of glory.[15]

Moltmann's resumption of Joachim of Fiore and Thomas Aquinas' conversation was motivated by a desire to affirm the place of chiliasm (otherwise known as millenarianism) within the Christian understanding of hope. He suggests that chiliasm is the immanent side of eschatology and eschatology the transcendent side of chiliasm. Interestingly Moltmann argues that one without the other has had disastrous consequences in European history. Without eschatology, chiliasm leads to political chiliasm. But, without chiliasm, eschatology is dissolved into transcendent longing, which has had equally disastrous effects.

Although he doesn't discuss it, Moltmann's balancing of chiliasm and eschatology potentially has positive ecological consequences. Chiliasm affirms *this* world as God's own, and prevents the sort of escapism so criticised by Keller and Ruether. Combined with an environmental ethic, this allows for ecological action, on the understanding that God rules over all of creation, not simply human history. But chiliasm is protected from destructive fundamentalism by the acceptance of eschatology as an integral part of Christian theology. Eschatology insists that the complete fulfilment of the kingdom of God has not yet occurred.

Part of the attraction of Moltmann's theology from an environmental point of view is his acceptance that eschatology can never be confined to personal eschatology. Moltmann's chief motivation in saying this is to develop a complete eschatology and so avoid the traps outlined by theologians such as Keller and Ruether. Critically, Moltmann argues that eternal life cannot be given or possessed by anyone without the fellowship of other people or the community of the whole creation. The temporal conditions of the whole of creation will be transformed, he asserts, and therefore cosmic eschatology is just as crucial as personal or historical eschatology. Moltmann makes explicit links throughout his work between

15. Moltmann, *History and the Triune God*, 96.

the various categories of eschatology. Therefore in his development of eschatological millenarianism he emphasises the strong links between the coming kingdom of Christ and the end of this world and the new creation of all things.

In an extremely helpful way, and in a similar way to Ruether, Moltmann highlights some of the dangerous examples of *historical* millenarianism, which he then contrasts with eschatological millenarianism. Of continuing interest to me is yet another theologian's criticism of the connection between millenarianism and fundamentalist apocalypticism in the United States of America. God's plan of salvation can be seen from the seven dispensations or periods to which the Bible witnesses. Moltmann here is critical of 'the great rapture', to which I referred in chapter one. An odd but intriguing debate has arisen within the 'dispensationalist' movement as to whether the rapture will save believers before the 'great tribulation', in the midst of it, or only afterwards. Within the American fundamentalist apocalyptic movement 'there are "pre-tribulationists", "mid-tribulationists" and "post-tribulationists"—for the initiated: "pretrib premils", "midtrib premils" and "posttrib premils"'.[16] Like Ruether and Körtner, Moltmann asserts that Ronald Reagan represented the linking of the political right with this apocalyptic fundamentalism. With Körtner, Moltmann identifies Jerry Falwell as another key American influence. Within the framework of millenarianism Moltmann then goes on to a detailed critique of American culture, which has elements of the 'Redeemer Nation' within its nation's key founding myths.

Having quite rightly dismissed the whole concept of historical millenarianism as an un-acceptable form of Christian eschatology, Moltmann then has the difficult task of defining how one might understand millenarianism. The key concept is that eschatology without millenarianism can only talk about a rupture of history, with the corresponding irrelevance for present ethics (a theme I shall discuss below). Consequently millenarianism is a fundamental part of eschatology, because the eschaton is the *consummation* of history, not its rupture.

Millenarian eschatology, if it is to be verified existentially, as opposed to being pure enthusiasm or a product of fantasy, must have a foundation in Christology. Here Moltmann's work on millenarianism touches base with the example of Christ, and he further develops his understanding of the new time of Christ. The coming of Christ also sees the dawning of the

16. Moltmann, *The Coming of God*, 159.

new, eternal aeon, which has come into the midst of this old aeon which is *passing away*. Here we are approaching something similar to Körtner's work on the unveiling of the true reality. The 'lordship' of the risen Christ is not yet fully manifested. So we also participate in Christ's struggle as we strive to follow in his footsteps, being a part of his messianic mission.

The dawning of this eternal aeon, which began specifically with the resurrection of Christ, means that the reign of Christ will come before the end of the world. There will be a time of transition, or, to be more accurate, a transitional kingdom, which will lead from 'this transitory world-time to the new world that is God's'.[17] This process of transition has already begun with the coming of Christ. But, in the general consummation of all things, or the general resurrection of the dead, as Moltmann talks about it here, the life of God's new world, already begun in Christ and experienced in the presence of his Spirit, will permeate every part of creation. What has already begun in Christ will be fulfilled completely, so that God will make all things new (see Revelation 21:5), or, as Moltmann later develops, God will be all in all.

One of the consequences of his theology is that Moltmann inevitably affirms a doctrine of universal salvation. For the purposes of this book, this doctrine is only of interest insofar as it pertains to all of creation. This is indeed what Moltmann works towards, particularly in his exploration of Christ's descent into hell. This is the ground for confidence that nothing will be lost. All of creation will be brought back again and gathered into God's eternal kingdom. From the profound suffering of Christ on the cross and his descent into hell comes the costliest grace, otherwise known as all-reconciling love. God will never give up that which he has created. Moltmann's focus here is on 'sinners, the wicked and the violent, the murderers and the children of Satan, the Devil and fallen angels',[18] whom, he insists, will be liberated and saved. Hence his focus here is anthropological. Nevertheless, as his section on cosmic eschatology[19] demonstrates, it is a legitimate extension from this position to include all of creation as participating in the universal restoration of all things.

Millenarian eschatology inevitably leads Moltmann to the question of time. I will discuss below in further detail some of the problems which arise for Moltmann and the question of time. Here, however, it is sufficient to note, consistent with Moltmann's understanding of *eschatological*

17. Moltmann, *The Coming of God*, 195.
18. Moltmann, *The Coming of God*, 255.
19. Moltmann, *The Coming of God*, 257–319.

rather than historical millenarianism, that time itself will be different. It is ridiculous to attempt to fit the coming of the messianic kingdom into calendar time. This is the mistake fundamentalist millenarianist movements have made consistently. God's coming kingdom, therefore, cannot and does not fit into any linear notion of time. Though Moltmann doesn't discuss it here, the fact that the kingdom is not connected to any concept of linear time also mitigates against the prediction of the parousia or the direct association of the parousia with particular historical events. Thus the various historical movements which have attempted to do so have misunderstood the essence of divine activity. As Moltmann puts it, 'the promises of the coming God are not surrendered to a so-called divine historical plan or "plan of salvation". They remain in God's hand.'[20]

It is here in this later work, *The Coming of God*, that Moltmann continues to explore the concept of different ages, which he developed in conversation with Joachim of Fiore. He suggests that, theologically, time is not determined by a linear concept, but rather by the different modes of God's presence in time. Just as we talk about different epochs, such the Middle Ages or the post-modern era, when talking about the changes that the human race has experienced in the whole paradigm of living, so too we can talk about different epochs of God's presence. For the Christian the present era, which Moltmann refers to as the 'kingdom of grace' is characterised by 'the presence of Christ in the life-giving Spirit'.[21] But we can also make distinctions between what Moltmann terms the 'kingdom of nature' where God is present as the creator and preserver, as well as the 'kingdom of glory'. This is where Moltmann directs his eschatology and, as I shall show, it forms the climax of his work. In this epoch, God himself will 'indwell his creation as in his temple'.[22]

His commitment to his particular project of reinvigorating millenarian eschatology leads Moltmann to affirm that millenarian expectation mediates between world history now and the end of the world and a new future to come. In other words there is a time of transition, because otherwise world history would simply end. This picture of eschatology, he suggests, leaves human beings with little reason to care for the way they live or act, because such action would be irrelevant for the end of the world. Such a theology would lead to irresponsible actions, simply because of the power of the mind-set which thinks, 'Eat, drink and be merry for tomorrow we

20. Moltmann, *The Coming of God*, 200.
21. Moltmann, *The Coming of God*, 200.
22. Moltmann, *The Coming of God*, 200.

die'. It leaves the coming generations no hope and leads the current generation to deluges, ecological, financial and nuclear. Future generations are then left to 'sink under the mountains of debt, atomic waste and the ravaged environment'.[23] The only viable form of eschatology, according to Moltmann, is a millenarian eschatology, because it is messianic, healing and saving.

In his development of millenarianism, Moltmann begins to develop an ecological eschatology, albeit without this being his specific intent. His work on the times of transition is helpful in the sense that it provides the reader with a reason for acting environmentally. If we are in a time of transition, then everything that we undertake which is positive, both in human society and the environment around us, will contribute to a smooth transition to the kingdom of God. This concept places a value on the work of human beings. It provides human beings with a proper basis for ethics and a powerful motivation for consistent discipleship of Christ. It allows the envisaging of a different way of being in the world, which often stands in stark contrast to the society around it. It is the foundation of every kind of Christian alternative form of life and action 'which will withstand the ravages of the world here and now'.[24] In a telling sentence, perhaps in response to the criticism of theologians such as Ruether, Moltmann insists that original Christian and Jewish millenarianism was a martyr eschatology, which is 'the precise opposite of every eschatological escapism'.[25] Moltmann's commitment to justice for the oppressed, including the environment, compels him towards millenarian eschatology, because it provides a strong basis for present ethics.

Just when Moltmann seems to have provided us with a form of eschatology which might give real hope for the eternal transformation of all things, and still allow an environmental ethic, we come to consider his work on evolution. In this respect Moltmann's work remains curiously inconsistent,[26] because the evolution of creation, even in the sense of evolving in a self-transcendent way, can contribute nothing to the even-

---

23. Moltmann, *The Coming of God*, 202.
24. Moltmann, *The Coming of God*, 201.
25. Moltmann, *The Coming of God*, 201.
26. See Douglas Schuurman, *Creation, Eschaton, and Ethics: The Ethical Significance of the Creation-Eschaton Relation in the Thought of Emil Brunner and Jürgen Moltmann* (New York: Peter Lang, 1991), 97. Schuurman makes the point that Moltmann's claims about the place of creation in the eschaton are internally inconsistent: 'On the one hand, Moltmann describes the eschaton in ways which deny the continuity with creation; on the other, he describes the eschaton in ways which demand continuity with creation.'

tual transformation of the world. This anomaly would seem to contradict his positive development of millenarian eschatology and the importance of the times of transition. It is to Moltmann's vexed relationship to the question of evolution that I now turn.

4.5 Evolution

One of the dilemmas which Moltmann faces, but, in the end, never resolves satisfactorily is the question of the place of the processes of creation, specifically evolution, in the final consummation of all things. This casts a question mark over his contribution to ecological eschatology and dampens an otherwise bright contribution in the form of millenarian eschatology.

As I indicated earlier, there are internal contradictions within Moltmann concerning his attitude to the processes of evolution. In his earlier work Moltmann attempts to comprehend and articulate a doctrine of creation, which is consonant with the accepted theory of evolution.[27] Using pneumatology as his starting point he develops a convincing understanding of creation itself being open to the movement of the Spirit, and creation being an open system, as opposed to one which is static and closed off to the future. Using a panentheistic model of creation, which is understood and described in trinitarian terms, Moltmann sees the created order as existing in God, and God dwelling in it. This trinitarian concept finds the balance between on the one hand a one-sided emphasis on God's transcendence, which can lead to deism, and, on the other, too great an emphasis on God's immanence, which leads to the pantheism of Spinoza. Moltmann correctly sees one of the challenges of contemporary theology as providing 'an integrating view of God and nature which will draw them both into the same vista'.[28] The critical distinction between God the Creator and the world as his creation gave human beings a sense of liberty, which they did not enjoy in the fertility cults. In these cults the divinisation of the creature determined, in a restrictive way, the relationship of the human being to nature. But once the critical distinction was made, the resulting freedom meant nature was regarded as totally godless, and so the relationship of human beings to creation became disastrous (though it's worth noting that the disenchantment of the world also was one of

27. See his work on 'Continuous Creation' in Moltmann, *God in Creation*, 206–214.
28. Moltmann, *God in Creation*, 98.

the motivators of modern science). Moltmann's work on the Trinity and panentheism gave him the model he wanted to integrate God's immanence and transcendence.

His resulting theology of the Cosmic Spirit and its role within creation and the processes of evolution is persuasive. It replaces what is in some Christian circles the dominant theology: a transcendent God ruling from on high, within a modern mechanistic worldview. This theology leaves nature both dead and spiritless, as well as godless. But Moltmann's work of the Spirit provides an alternative model. Rather than complex systems being reduced to simple ones to explain their complexity, as tends to occur within the reductionistic framework of the modern mentality, Moltmann's pneumatology suggests the opposite. Complex systems explain the simpler ones. Thus Moltmann uses human experiences of God and complex relationships to God to draw certain conclusions about human and natural conditions. Starting from that principle, Moltmann draws conclusions about the way the cosmic Spirit operates in nature.

For example, the Holy Spirit is viewed by Moltmann as permeating all of creation and dwelling in every individual creature and the community of creation.[29] The Spirit is involved in the processes of evolution at a fundamental level, creating interactions, harmony and mutual perichoreses and promoting a life of co-operation and community. In this way all created things are manifestations of the divine Spirit, because everything is representative, even if in an incomplete way, of the infinite. Moreover, the presence of the Spirit in creation—the infinite in the finite—means the finite creation is imbued with self-transcendence. Because of their openness to the Spirit, all creatures are 'aligned towards their potentialities'.[30]

Moltmann engages in a helpful exegesis of Romans chapter 8 in his discussion of the notion of self-transcendence. He suggests that the focus of Paul's writing here is the term 'longing' or 'yearning', which indicates a desire of the whole created order to be redeemed with the children of God (the resurrected believers). Moreover, the believers themselves are drawn into solidarity with other created things. The presence of the Spirit in creation is a sign of the self-limitation of God. God is not present in creation through the Spirit as an omnipresent ruler, otherwise the infinite would subsume the finite. There is a self-humiliation of the Spirit, a divine *kenosis*, which allows the Spirit to dwell within creation without dominating it. In the sufferings of creation, including the aberrations of evolution and

29. See Moltmann, *God in Creation*, 100.
30. Moltmann, *God in Creation*, 100.

the whole history of human crimes and catastrophes, the Spirit is present, suffering with creation, longing for the transformation of all things. In the suffering of creation and human beings and their cries to God, we may also discern the yearning of the Spirit, who at the same time imbues creation with a yearning for self-transcendence.

The combination of panentheism with trinitarian theology is a powerful one for Moltmann, who suggests that this is the only way to comprehend the differentiation between God's distinctiveness from creation and his presence in it. It is precisely language about the third person of the Trinity which allows him to achieve the balance between immanence and transcendence. It is the Spirit who differentiates and yet binds together. 'The Spirit preserves and leads living things and their communities beyond themselves.'[31] It is the cosmic Spirit, which Moltmann views as one and the same as the Holy Spirit, the third person of the Trinity, who enables the entire creation to be an open system, that is, open to God and his future.

At this stage in his work Moltmann seems to have an overwhelmingly positive view of creation. God is present throughout creation through his Spirit. 'The whole creation is a fabric woven and shot through by the efficacies of the Spirit.'[32] Indeed creation moves and evolves in the 'energies and powers of the divine Spirit'.[33] Moltmann is alluding here to the Orthodox doctrine of essence and energies.[34] God the Spirit is present throughout the evolving world, encouraging self-transcendence, provoking new paths of evolution, through the Spirit's uncreated and creative energies.

Moltmann's pneumatology finds its fullest expression in his development of the doctrine of creation, especially *creatio continua*. It is here, however, that we begin to see a distinction made between the various forms of creation, namely creation in the beginning, continuous creation and the consummation of creation. It is these distinctions which contribute to Moltmann's later rejection of much that is positive about evolution.

Nevertheless in his earlier work, *God in Creation*, he still views evolution in a positive light. God created the world out of nothing, and this doctrine of *creatio ex nihilo* expresses the miracle of the world's existence, as well as its contingence. Having created the world, however, God did not

---

31. Moltmann, *God in Creation*, 103.
32. Moltmann, *God in Creation*, 212.
33. Moltmann, *God in Creation*, 212.
34. See Duncan Reid, *Energies of the Spirit. Trinitarian Models in Eastern Orthodox and Western Theology* (Atlanta, Georgia: Scholars Press, 1997).

let it stand alone or remain static, but established the stage of evolution as his free creations, with the result that creation was open to the future. It was never a closed system. Through his Spirit God ensures that the creation is given movement and direction, and he opens up possibilities and entices creation in the direction of these possibilities. This openness of creation from the very beginning means the conditions were established for creation's history of 'perdition and deliverance, *as well as its consummation*'[35] [my italics].

The point at which Moltmann's theology of evolution begins to be less helpful, in terms of developing an ecological eschatology, is when he explores the notion that *creatio continua* does not produce any new contingencies. It simply means the continual preservation of what God has created and the preparation for its completion. The conditions are present for creation's consummation, but *creatio continua* does not bring this consummation about. Thus Moltmann insists on a threefold process of creation, which, whilst obviously connected, are also differentiated at a basic level. Using the Latin terms, the three stages of creation are: *creatio originalis*, *creatio continua* and *creatio nova*. Thus God's creative activity in history and time, through the Spirit, only anticipates the ultimate consummation of all things. It is a process of preparation for the new creation.

Yet even at this point Moltmann's attitude to evolution and the processes of creation (understood in the doctrine of *creatio continua*) is still positive. He is at pains to link the three stages of creation, more than to emphasise their separateness:

> So it is theologically inadequate if we restrict the divine creative activity to the beginning, and in a historical context talk only about 'preservation', and eschatologically only about 'redemption'.[36]

Clearly Moltmann links God's creative activity with his continual preservation of that creation. The preservation of creation occurs precisely through God's continual involvement in creation, through the Spirit, in the creative processes, namely evolution. There is also a clear link between God's preservation and God's consummation of creation, and for this Moltmann uses the language of initiation: 'The revelation of the rule of God in history *initiates* the fulfilment of that promise which creation

35. Moltmann, *God in Creation*, 207.
36. Moltmann, *God in Creation*, 208.

at the beginning represents in its very self'[37] [my italics]. But Moltmann vacillates between the language of *initiation* and *preparation*.

In the end, though, Moltmann, concludes that the consummation of creation cannot be thought of as the final conclusion and completion of the processes of an open system. Rather this consummation will occur through the indwelling of God himself, who will drive out the forces of negativity, as well as fear and the struggle for existence. Yet even here Moltmann views the indwelling of God as the *completion* of the processes of creation. Thus he still views these very processes in a highly positive light. Admittedly it is God in his transcendence who brings all things to completion, but it is the completion of an open system, which God has been intimately involved with from the very beginning.

Moltmann's views on evolution are not consistent, however, and appear to change in the years between publications. Perhaps it all begins with sin! For Moltmann, questions of finitude, of death, and particularly of sin loom large in his theological thinking. As a result we see Moltmann's attitude to evolution change as his own thinking progresses and so a contradiction appears in Moltmann's collective theology. Whereas in his earlier work, *God in Creation*, Moltmann had seen a positive role for the Spirit in the processes of evolution, with these processes themselves being viewed in a highly positive way, his later work is quite different. Here Moltmann views evolution as something from which all of creation needs to be redeemed by Christ. The contrast is stark and represents something of a difference between Moltmann's pneumatology and his Christology. In his later work even the language used changes. No longer is evolution portrayed in the overwhelmingly positive light that it is in *God in Creation*. Suddenly Moltmann talks about catastrophic breakdowns in the process of evolution and, using even more emotive language, about evolution's 'victims'. Whether the language is appropriate is an interesting question. The end of one dominant species often allowed another to flourish, including the human species.

Moltmann begins his work on evolution in this later book on Christology by suggesting that the divine creative and redemptive activity within human history has hidden correspondences in the world of nature. He admits that we can discover 'even in the history of nature parables and true symbols for the future of creation in its completion and perfecting'.[38] But it is interesting to note that even in this still positively phrased in-

37. Moltmann, *God in Creation*, 209.
38. Moltmann, *The Way of Jesus Christ*, 291.

troduction to the topic the language of initiation has been discarded in favour of the language of parables, symbols and preparation. What we see in the development and progress of nature is God *preparing* the world for its consummation. He reminds us, though, that God is already present in creation through the Spirit.

Fairly quickly, though, Moltmann begins to critique the theology of those who see evolution and its processes in a positive light in terms of salvation. He examines the work of both Teilhard de Chardin and Karl Rahner. Teilhard's main contribution, as Moltmann sees it, is that the increasing consciousness of life through the processes of evolution is leading to a process of divinisation. The person of Christ is the beginning of the divinisation of humanity, which brings with it the deification of the cosmos; 'for in this cosmic Christ the becoming human of the universe and the becoming human of God converge'.[39] Teilhard used the Eucharist as his inspiration. Just as the bread is transformed in the eucharistic mystery into the body of Christ, so too through his understanding of christological evolution 'God becomes part of the world and the world is deified'.[40] It is a vision of a cosmic Eucharist.

Moltmann's criticism of Teilhard is that Teilhard simply doesn't take account of the failures within the evolutionary process, due to the sacrifice of many living things to the survival of the fittest. Indubitably, within the processes of evolution, increasingly complex life systems do develop, but 'in the same process milliards of living things fall by the wayside and disappear into evolution's rubbish bin'.[41] Moltmann highlights here the cruelty of nature, because of its rejection of the weak, sick and unfit. He suggests that any concept of universal reconciliation is contradictory to the whole thrust of evolution. A critical question of Moltmann at this point is whether the language of cruelty as he applies it to nature or evolution is really appropriate.

Moltmann concludes his criticism of Teilhard by suggesting that his Christology is the weak point. He maintains that Teilhard's understanding of a *Christus evolutor* without *Christus redemptor* 'is nothing other than a cruel, unfeeling *Christus selector*',[42] because such a Christ is uninterested in the victims of evolution nor has compassion for the weak, who struggle for survival. Moltmann's key point is that the perfection of creation is

---

39. Moltmann, *The Way of Jesus Christ*, 294.
40. Moltmann, *The Way of Jesus Christ*, 294.
41. Moltmann, *The Way of Jesus Christ*, 294.
42. Moltmann, *The Way of Jesus Christ*, 294.

simply not achievable in any way through the processes of evolution or self-transcendence. Christ is the redeemer of evolution, the saviour of the victims, the one who suffers with the victims and so redeems them. The victims of evolution experience no joy through the processes of evolution, but only in the resurrection of nature itself. Evolution is, Moltmann, declares, too ambiguous and has no redemptive efficacy and therefore no salvific significance. Moltmann suggests that if Christ is to be thought of in conjunction with evolution he must be thought of as evolution's redeemer.

Thus we see in the language and concept that Moltmann's thoughts about evolution and indeed creation itself are far less positive in this later work. Almost but not quite absent is the beautiful language of the cosmic Spirit, present in all of creation with his uncreated and creative energies. In its place is an ambivalence towards the progress of creation and an emphasis on 'dying, violent death, mass extermination and the extinction of whole species through natural catastrophes and epidemics'.[43] Every step of progress in the processes of evolution, Moltmann reminds us, has its victims, those who did not survive. In what seems a direct contradiction of the language of *God in Creation*, Moltmann insists that 'the various processes of evolution in nature and humanity can only be brought into a positive relationship to Christ, the perfecter of creation, if Christ is perceived as a victim among evolution's other victims'.[44] Forgotten is Moltmann's own work on the presence of the divine Spirit, encouraging and provoking creation in its process of self-transcendence.[45] The difference between the two approaches is as remarkable as it is inconsistent.

Moltmann's rejection of evolution continues in his critique of Rahner. Moltmann chooses to critique Rahner's use of the concept of self-transcendence, at least as it pertains to the consummation of all things. Briefly, Rahner's concept of Christology within an evolutionary framework suggests that the Spirit, dwelling in nature, drives creation towards God in a process of self-transcendence. This self-transcendence already involves God's communication of himself. The climax of this process of self-transcendence is reached in the person of Christ.[46] Thus the peak of self-tran-

43. Moltmann, *The Way of Jesus Christ*, 294.
44. Moltmann, *The Way of Jesus Christ*, 294.
45. See Moltmann, *God in Creation*, 101.
46. Karl Rahner, 'Christology within an Evolutionary View of the World', in *Foundations of Christian Faith: An Introduction to the Idea of Christianity* (New York: Crossroad, 1978), 195.

scendence and divine self-communication occur in the Logos, in whom God communicates himself completely. By implication the earlier stages of human history, particularly through the process of self-transcendence, lead up to this moment of complete self-communication by God.

Moltmann's main criticism of Rahner is the same as of Teilhard. The ontology of self-transcendence within the framework of evolution sees Christ as the summit of development, 'but not as the redeemer of that development from its ambiguities'.[47] Christ therefore is the climax of self-transcendencies and opens up a new divine history in the world. But Moltmann suggests that this Christ does not stand beside 'the self-transcendencies which have miscarried and been thwarted. It is hard to see what redemption this Christ can bring to the graveyards of nature and human history.'[48] Moltmann seems at this point to be unaware of Rahner's later work on the topic. In fact Rahner did address some of the problems thrown up by evolution and suggested that one must allow for surprises and dead ends, but that a crucial affirmation of the Christian faith is that a total halt to progress will never occur.[49] However, Moltmann makes no reference to the article.

It's worth noting that Moltmann does accept and take up Rahner's notion of self-transcendence in a particular way, namely as it pertains to *creatio continua*. He acknowledges that active self-transcendence '*really does point* beyond the historical present and beyond history itself to a future which fills it and brings it to rest'[50] [my italics]. However, in the end he rejects the evolutionary processes of the world as having anything to contribute to the understanding of the *consummation* of all things. The deification of the cosmos is only understandable if its basis is in the cosmic incarnation of God. Evolution leaves too many victims in its wake. Even if the lower life forms can be understood as still being inherent in the high life forms, nevertheless they are not present as individual examples. This does not lead to their immortality. Whilst a perfect higher being is considered the possible end of the processes of evolution, this is not the eschatological consummation of all things.

Moltmann therefore sees redemption as in fact running *counter* to evolution. His reversal from his earlier work in *God in Creation* is complete.

---

47. Moltmann, *The Way of Jesus Christ*, 300.
48. Moltmann, *The Way of Jesus Christ*, 300.
49. See Karl Rahner, 'Natural Science and Reasonable Faith', in *Theological Investigations XXI* (London: Darton, Longman & Todd, 1988), 1–55.
50. Moltmann, *The Way of Jesus Christ*, 304.

The key issue for Moltmann, as it is in all of his eschatology, is time. It is from the future that God's redemption comes—salvation does not run in a teleological sense from the past to the future in the processes of evolution. The redemption which the Day of the Lord brings is the consummation of all things. All things will be gathered together and nothing will be forgotten; every last created being will be gathered into God's presence. This is a vision which evolution cannot achieve. Indeed evolution itself will be redeemed from its own ambiguities. Thus the redemption and the new creation can only be expected from '*the coming of Christ* in glory'.[51] The processes of self-transcendence, then, do point beyond the present to the future, but they do not contribute to the realisation of the consummation of all things. They are 'at best a glimmer of hope for this unredeemed world'.[52]

Moltmann's rejection of the processes of self-transcendence or evolution as having any role, even participatory, in the final consummation of all things leaves his eschatology open to ecological criticism. If the very processes of the world around us have no lasting contribution, in the sense of contributing to the consummation of all things, are they really important at all? Using personal language we might phrase the question as, 'Is my life's work at all significant, if it contributes nothing, in the end, to God's purposes for humankind and the cosmos?' Do we really need to care about the environment, if the environment actually contributes nothing of lasting value to the fulfilment of creation in the eschatological consummation of all things? Moltmann does not resolve these questions in his work on evolution and its place in the redemption of the cosmos.

4.6 Apocalyptic eschatology

Moltmann's rejection of self-transcendence or evolutionary processes even contributing to the final consummation of all things leads him to the question of apocalypticism. If the cosmos cannot arrive at a state of fulfilment within itself (whatever role God might have in the process of self-transcendence) then the end of the world must come in order to facilitate the transition to the new creation.

Moltmann begins his examination with a helpful analysis of the various catastrophes facing the world at the moment, including the potential

51. Moltmann, *The Way of Jesus Christ*, 304.
52. Moltmann, *The Way of Jesus Christ*, 305.

for nuclear annihilation, the biological and genetic destruction through the release of bacteria (one could add here viruses as well), ecological disaster, and the economic enslavement of the developing nations. He is helpful, particularly in the light of American fundamentalist apocalypticism, in pointing out that any such end of humanity and indeed the world would be the result of human beings' actions, not God's actions. But it is precisely God who would initiate the end of the world, with the promise of a new world to come, at least according to positive apocalyptic models, to use Körtner's apocalyptic framework. Moltmann, like Körtner, reminds us here of the true meaning of apocalypse, which, linguistically at least, has nothing to do with the end of the world *per se*, but rather means unveiling, making manifest or disclosing. The threat of nuclear destruction, or, to use a current example, the heightening of tension in the Middle East including Israel, has nothing to do with true apocalypticism. Indeed such philosophising pushes the responsibility of human action on to God. 'That is the height of godlessness and irresponsibility. The first work of true apocalyptic eschatology must be to expose the mystifying, deluding use of apocalyptic language at the present time.'[53] Moltmann is extremely helpful both in a political sense and an environmental one in his exposure of false (or damaged) apocalypticism. He refers to historical apocalypticism, as represented by the kinds of human-caused destruction referred to above, and contrasts this with apocalyptic eschatology or millenarian apocalypticism. I think Körtner's description of damaged apocalypticism is probably more helpful, because it highlights the suppression of apocalypticism and its subsequent damaging at the hands of fringe groups. But whichever term we decide to use, there is no doubt the association of the modern state of Israel with some kind of catastrophic end of the world, and the location of the new Jerusalem, is a powerful and persistent myth. Moltmann and Körtner's exposure of this kind of thinking can only be helpful, as theologians attempt to guide people back to true Christian apocalypticism, or what Moltmann terms apocalyptic eschatology[54] and Körtner the sublation of apocalypticism.[55]

53. Moltmann, *The Coming of God*, 218.

54. Moltmann helpfully points out that 'apocalypticism belongs to eschatology, not to history. And yet eschatology begins with apocalypticism: there is no beginning of a new world without the end of this old one'. *Coming of God*, 227.

55. Körtner's concept of the sublation of apocalypticism is that Christianity is not simply a variation within apocalypticism. Apocalyptic world anxiety is part of the disposition of those who also hope for deliverance. But this deliverance is seen precisely in the end of the world, which is seen positively, because it overcomes the inescapability of

The vexed question of the relationship between apocalypticism and history is one which Moltmann then confronts. This is an issue closely related to the issue of time, with which Moltmann struggles continually. If, as Moltmann does, we reject any concept of human beings somehow being responsible for apocalyptic destruction, and if we reject cosmic catastrophes as being God's judgement on the world, where and when does the apocalypse occur? Central to all of Moltmann's eschatology is the assertion that apocalypticism belongs to eschatology and not to history. The 'Day of the Lord' is not an event in history, as Moltmann will later assert when discussing time. I will discuss this issue more fully below.

It is in discussing the possibility of the apocalypse that Moltmann begins to stumble, because, as Körtner suggests, he is unwilling to face the concept of total destruction as a possibility within his theological schema. Despite Moltmann's affirmation of the apocalyptic nature of eschatology, and despite his confronting of the questions of theodicy, he nevertheless evades the idea of an end whose actuality has become possible. Körtner questions whether Moltmann's theology is prepared to confront, without reservation, the possibility of total catastrophe.[56] This is despite Moltmann affirming that Christology provides apocalypticism with its model. 'The raising of Christ from the dead presupposes his real and total death. It is from this fact that Christian apocalyptic takes its bearings: his real end was his true beginning.'[57] Here we begin to see another of Moltmann's internal contradictions. At this point in his work he affirms that the whole cosmos will participate in Christ's tribulation, 'so that it may be drawn into the cosmic resurrection and new creation.' He affirms the place of apocalypticism within the Christian tradition, in the sense that the world and the cosmos are drawn into the Christ-event. Using Paul's imagery from the book of Romans (chapter 8), Moltmann suggests that we are all suffering, human beings and the wider creation alike, so that we may be drawn into Christ's resurrection. Thus the whole of creation will experience cosmic resurrection and become a new creation. At a cosmic level this reflects what Moltmann confirms about the Christian journey as well, in that Christians experience Christ's death and resurrection through the symbolism of baptism (again following Paul's theology in Romans chapter

finitude. Thus whilst faith stands within the framework of apocalyptic world anxiety, it also believes that the fate of Jesus has broken through the structure of reality disclosed by that anxiety.

56. See Körtner, *The End of the World*, 264.
57. Moltmann, *The Coming of God*, 227.

8). It is out of the universal suffering in the cosmos that we then gain hope from the visions of the coming kingdom of God.

At this point at least Moltmann belongs to the tradition of positive apocalypticists, who see hope in the end. He maintains, quite correctly, that apocalypticism preserves hope from falling victim to the empty promises of facile optimism:

> Eschatology is not a doctrine about the world's happy end. In the present situation of our world, facile consolation is as fatal as melancholy hopelessness. No one can assure us that the worst will not happen. According to all the laws of experience: it will. We can only trust that even the end of the world hides a new beginning if we trust the God who calls into being the things that are not, and out of death creates new life.[58]

It is out of the destruction of the world that true hope emerges, or at least *may* arise. We look beyond the horizon of this world's end to the emergence of God's new world. Interestingly Moltmann suggests that this hope in the face of the end encourages us to an *act* of hope. The coming world of righteousness, justice and peace inspires us to strive for such things now, contrary to appearances and to all apparent chance of historical success. This is because we say 'yes' to life, even in the face of impending destruction.

But whether Moltmann is truly convincing at this point is debatable. In the face of the end, particularly given the apparently limited nature of our contributions, are we in fact inspired to act with justice, including justice for the environment? Moltmann fails at this point to find the balance between negative and positive apocalypticism, to use Körtner's terminology again, and ends up simply asserting the importance of acts of righteousness, without providing the theological impetus to inspire them. In fact, as we shall see, the eschatological models Moltmann proposes move away from the concept of apocalyptic eschatology and towards something else: Shekinah theology. This is actually a positive move for Moltmann in terms of ecological action. But it does reveal to us the limits of Moltmann's apocalyptic eschatology. The possibility that absolute destruction might occur is not dealt with, but rather simply rejected as a faulty eschatological model.

58. Moltmann, *The Coming of God*, 234.

Moltmann's rejection of absolute destruction rests on his definition of apocalypticism. He suggests that the Lutheran understanding of the annihilation of the world is not apocalypticism as such, but rather exterminism. Instead of transformation or the coming of God into creation, we are rather left with the movement of being to non-being, the reverse of creation itself. Part of the criticism that Moltmann levels against this particular strain of Lutheran theology is that there is no suggestion of the creation of a new world. Rather the blessed will no longer require the created surroundings of heaven and earth, because they will live in God and God in them. God then becomes their environment and there is no need for the world at all, hence its utter annihilation. The earth becomes irrelevant in this particular theological schema. Human beings are absorbed wholly into the beatific vision. Moltmann quite correctly holds to the view that the concept of resurrection is lost 'if salvation is supposed to consist only of the blissful beatific vision of the disembodied soul'.[59] He suggests that this particular strain of Lutheran theology then potentially becomes a kind of world-denying and world-despising Gnosticism. He rejects it totally and notes, in passing, that most modern Lutheran theologians have reverted to the more traditional understanding of the transformation of the world, not its annihilation.

If we agree with Moltmann's rejection of the concept of annihilation, as developed by some Lutheran theologians in the past, we are still left with the issue of the utter destruction of the world. Despite his earlier affirmation of the end of the world, as developed in his apocalyptic eschatology, he never addresses what it might mean if the world were totally destroyed, only then to be created anew. Having rejected the concept of world annihilation, as developed by some Lutheran theologians, for example Paul Gerhardt,[60] he then begins to use the quite different language of *transformation*. It is a telling shift in language and lends credence to Körtner's view that Moltmann never really deals with the possibility of the utter destruction of the world.

However, despite the contradictions within Moltmann's work, his eventual model of God's Shekinah does have something overwhelmingly positive to contribute to the development of ecological eschatology.

59. Moltmann, *The Coming of God*, 270.

60. Moltmann, *The Coming of God*, 269.

4.7 God's Shekinah

Before we come to Moltmann's development of God's Shekinah as his principal model of eschatology, it is worth examining the analysis he makes of various other eschatological models, because this provides us with an insight into his own thinking. For Moltmann the approach taken by Calvinist theologians of the seventeenth century provides him with more inspiration than the approach of Lutherans of the same century. It is from the Calvinists that he acquires the language of transformation. The Calvinists did use the language of annihilation as well, but only in the context of God annihilating the world in its *present form*. Out of this world, which will have been judged (and presumably purified) by fire, God will make a new world. But this new world will retain its identity as God's creation, rather than something quite different. Thus, in the Calvinist theology of the seventeenth century, the new creation enjoys both continuity and discontinuity with the present created order. The example for this is imagined in the eternal life of every believer, who grows into the fullness of life in God, out of the *seeds* of eternal life which were a part of the believer in his or her earthly life. So too the new creation comes from the old, in a process of transformation from transience into eternity.

However, Moltmann suggests that the language of transformation does not go far enough, and proposes instead that the eschatological transformation of the world means 'a fundamental transformation, that is to say a transformation in the transcendental conditions of the world itself, and therefore of its very foundation: God himself changes his relationship to the world'.[61]

This language of fundamental transformation might at first appear to bring Moltmann back closer to apocalypticism, but in fact it simply leads him to explore exactly how God does change his relationship to the world. The concept of God's Shekinah arriving to dwell in creation is Moltmann's answer.

Moltmann's next step along the road to his development of a theology of God's Shekinah is an examination of Orthodox theology and its idea of the deification of the world. This concept initially appears to be anthropologically centred, because it is referring to the transformation of human beings, that they share the characteristics of the divine nature, through their community with Christ. It implies a sense of kinship with God, specifically with respect to those characteristics which are salvific for human

61. Moltmann, *The Coming of God*, 272.

beings, that is, non-transience and immortality. It is the Orthodox development of Paul's concept of divine sonship and daughtership, which is ascribed to believers. We become heirs with Christ of the promises of God.

Moltmann notes that this human-centred doctrine was then extended to become a cosmologically-centred one. This occurred easily within Orthodox theology, because of the less rigorous distinction made between person and nature, as made in Western theological traditions. There is a hypostatic bond between all human persons and nature itself. It follows from this bond, that if human beings are redeemed that is, deified, then all of creation is as well. 'The redemption of human beings draws in its wake the redemption of nature.'[62] The key link is the fact that we are embodied creatures. Through the transfiguration of human bodiliness, as disclosed, for example, in the Transfiguration of Jesus on the Mount, the whole of nature is gathered into the fellowship of this transfigured, transformed humanity.

The Eastern Orthodox doctrine remains human centred, but this need not be a fault, if the intimate connection between human beings and creation is always remembered. Anthropocentrism only seems to be a potential weakness, in an ecological sense, in the Western tradition. In the Orthodox tradition the connection between human beings and the rest of nature means that even an anthropocentric doctrine such as the deification of human beings need not result in an abandonment of the importance of creation itself.

Moltmann, however, goes further than the Orthodox tradition and extends this doctrine to include the idea that the whole of the cosmos is transformed into the likeness of God: 'the whole redeemed creation enters into the light of God's truth and is deified'.[63]

The concept of human beings as priests of creation is, unfortunately, one which Moltmann doesn't develop. This would actually supply him with a means to link human beings' actions with God's transformation of the cosmos. This sense of us being part of the community of all creation, with human beings playing a crucial role as spiritual beings in bringing creation to God, is one I will explore in a later chapter.

Moltmann does identify one problem with the Orthodox doctrine of the deification of the cosmos, which is that it is not really the creation of a new heaven and earth. Rather the deification of the cosmos is seen as the spiritualisation of the cosmos and the interpenetration of the radiance of

62. Moltmann, *The Coming of God*, 273.
63. Moltmann, *The Coming of God*, 273.

the Spirit. He suggests that this leaves Orthodox churches and their spirituality open to an element of docetism. He notes that the real problem with this Orthodox doctrine, from a christological perspective, is that it corresponds to a one sided theology of the resurrection. However, combined with his development of a Christian Shekinah theology, Moltmann successfully uses the Orthodox doctrine as the basis for his understanding of the coming kingdom of God.

Moltmann's eschatology reaches a high point, here, when he suggests that there is no salvation without the world. The critical question then becomes whether the world will be a redeemed world, after this contingent and transient world has come to an end, or whether we should return to the promises of an earthly future, as prophesied by Israel's prophets.

Having spent so long advocating the merits of apocalyptic eschatology, even his own work doesn't *truly* belong to that category (because the concept of total destruction is rejected), it should come as no surprise that Moltmann rejects those positions which return to the promises of an earthly future. It is here that we see Moltmann engage with eco-feminism for the first time in his work on eschatology. He correctly identifies eco-feminist theology, particularly Ruether's, as in fact returning to a thoroughly prophetic understanding of the future promises of God. As I discussed in chapter one, the central argument of some eco-feminist writers, an argument which Moltmann also examines, is that the yearning for some kind of immortal existence is actually hostile to life itself. Instead we should concentrate on the kinds of prophetic visions of the Old Testament and strive to live in the golden ages promised. Thus the vision which all Christians and indeed all people should strive towards is that of old people who live out their days, and communities which build their own houses and live in them and plant their own vineyards and eat their fruit. Death itself is not something to be feared, but rather welcomed, in old age at least, as part of the natural cycle of life. Whilst we must protest against early or unnatural death, death itself is part of the natural process, in which the older generation passes away and a new one takes its place. For our part, when we approach old age, we welcome death as a time of returning to the wider matrix of life. So we transcend our egos and the desire for immortality and instead recognise our part in the wider community. We are buried in the earth and so become compost for the future, being, as it were, reabsorbed into the great organism of all living things, otherwise known as *Gaia*.

Having spent chapters one and two critiquing exactly this theology as it's developed by Keller and Ruether, I will not repeat myself here. It is, however, interesting to pick up on two of Moltmann's criticisms of some eco-feminist theology. Firstly, the underlying assumption of this approach is that the world is both good and everlasting. The assumption that the world is good is an interesting one. As I shall show in chapter six, Ernst Conradie has shown that one part of the earth's redemption is redemption from sin. Amongst certain theological movements (and eco-feminism is one) the reality of sin is downplayed. There is also an assumption that in this golden age, promised by the prophets, this good earth will 'simply have to flower into a new undreamed of fertility'.[64] The underlying assumption here is that the cosmos itself will never be threatened. But this overlooks that fact that the earth is remarkably fragile and destructible. It assumes that the fruitful cycles of nature will *always* continue. It ignores the scientific theories of entropy. It also ignores the possibility of the earth experiencing another ice age, and all human life ceasing. It ignores the scientific prediction that this earth will eventually melt in the fierce heat of the sun as it expands to become a red giant. In summary this particular eco-feminist vision fails because, unlike the writers of the Old Testament prophetic literature who had no such scientific information, it ignores the reality that the earth, and indeed all of creation, is contingent and not eternal or everlasting.

Moltmann's second criticism of interest is that the vision of some eco-feminist theologians assumes that hope for resurrection is necessarily at odds with respect for creation. But, as Moltmann points out, the Orthodox hypostatic link between human beings and the rest of creation means that human beings long to be redeemed *with* this creation. Just as it makes no sense to reject the earth and leave the redeemed without a place to be embodied, albeit in their resurrected state, neither does it make any sense for the body to be dissolved in the earth. Just as the hypostatic unity emphasises the connection through the human body between all human persons and the entire creation so too it differentiates between the two. If this differentiation is resolved in the favour of one side or the other then there is no redemption.

So we come to Moltmann's own vision of eschatological ecology, as he terms it. It builds on the work of one of his predecessors at Tübingen, Johann Tobias Beck (1804–78). Beck's vision is not of a restored earth,

64. Moltmann, *The Coming of God*, 275.

in the sense of being restored to some kind of original paradise, rather, much like the Orthodox doctrine of deification, the entire world becomes shot through with the divine Spirit. It becomes a new world-organism, in which heaven and earth are united. God and human beings too then dwell in perpetual fellowship. The term which Beck uses to describe this new world-organism is the christological idea of perichoresis, which Moltmann himself has also used in his writings on the Trinity.[65] The earthly sphere is no longer divided from the heavenly one, but rather, through this perichoretic movement, or mutual interpenetration, God is all in all. Fellowship in God and fellowship in the world are then no longer mutually exclusive or distinctive. The model for this, according to Beck, which Moltmann finds so fruitful, is the doctrine of the two natures of Christ (divine and human). Both are possible within the person of Christ, because of the idea of *perichoresis*. This then becomes the model for the 'mutual indwelling' (the term Moltmann settles on) of earth and heaven.

Here Moltmann's eschatology takes a decidedly positive ecological flavour, because this new world comes from this current one. This temporal earth holds within itself the promise of the new earth, in which God will dwell. The earth itself becomes indeed the bearer of our future. Fellowship with Christ and with God is inextricably linked with fellowship with the earth. There is not one without the other. Moltmann then finds the balance between the rejection of the earth as expendable, dead matter, which we exploit, and the eternal goddess Gaia. Neither model is satisfactory. The first leaves the earth vulnerable to all sorts of exploitation, which the human race has managed with alarming success. The second model in fact deifies the earth itself, attributing eternal, divine characteristics to the contingent creation, surely a definition of idolatry! But Moltmann's model leaves us with a way forward. The earth is valued as the bearer of the promise of something greater. But ultimately the earth itself does not have to, and indeed cannot, provide its own redemption, or ours, rather redemption comes from the mutual indwelling of God and the earth. Nevertheless we look to the earth as giving us a real experience of the promise to come and as being the locus of that promise.

65. Jürgen Moltmann, *The Trinity and the Kingdom of God: The doctrine of God* (London: SCM, 1981), 174.

4.8 Space and God's Shekinah

In order to understand how this concept of mutual indwelling might function in relation to the idea of God's Shekinah dwelling amongst us, it is necessary to evaluate Moltmann's theology of the space of creation. Fundamental to this is his development of the idea of God restricting God's self in order to give creation space to come into being. A similar concept is used by Moltmann in relation to time: 'God restricts his eternity in order to take time for his creation and to leave it time'.[66] In the same way God's self restriction of his omnipresence allows creation room to grow. God withdraws himself in order to allow room for something else. But, of necessity, from this self-restriction emerges the distance between God and his creation.

But, as Moltmann goes on, the distance is not so great as the idea might at first imply. God makes room for creation within himself. It is in fact quite a feminine image of God, as the God who makes a place for a growing creation. The Creator then becomes the God who is inhabited. The concept depends on Moltmann's perichoretic understanding of the Trinity. The three persons of the Trinity dwell within one another, in a reciprocal in-existence, as he terms it.[67] Thus the doctrine of the Trinity does not refer simply to the relationship of the three persons to one another, but also their reciprocal indwellings. They exist mutually in one another, thus giving one another space for a full unfolding. Within the Trinity space is then made for creation. This is a development of the statements in John's gospel, in Jesus' high priestly prayer (see John 17:21). The divine persons open themselves 'to become the living space for created beings'.[68]

Moltmann defines this as a perichoretic concept of space: space as mutual indwelling. The earthly example of this *par excellence* is of course the experience of the foetus in the uterus. This is a good model of how God makes room within God's self for creation. The child grows within the mother's womb, depends utterly on the mother for its existence, and yet is distinctive in itself and has its own space. But in fact every living creature lives in community with other living creatures. Nothing in the world can exist by itself, just as empty space cannot exist either without objects to rest or move in it. The perichoretic concept of mutual habitation provides

66. Moltmann, *The Coming of God*, 297.
67. Moltmann, *The Coming of God*, 298.
68. Moltmann, *The Coming of God*, 299.

us with a good ecological understanding of the interrelationship of all living things.

This brings Moltmann to the question of how God relates to his creation, specifically how he dwells within creation, if he has restricted himself in order to give creation space. How can God dwell within the earthly spaces without destroying these limited spaces through his infinity? The question leads Moltmann to arguably his most interesting contribution to eschatology in particular and theology in general: his development of the doctrine of the Shekinah in a Christian eschatological context.

The concept of the Shekinah 'links the infinite God with a finite, earthly space in which he desires to live. Shekinah theology is temple theology.'[69] It describes God's desire to be in a particular place, based on an act of descent and self-humbling. The Christian equivalent is the *kenosis* (self-emptying) of the Logos in the incarnation. There is, then, a sense in Shekinah theology, just as there is in incarnational theology, of a self-differentiation of God, that is, of God who reigns in heaven and God who lives in the midst of the people. The idea of the Shekinah included the sense of the Shekinah journeying with the people, and suffering with them. There is also the understanding that there are times when the Shekinah is close by the people and times when it is quite remote. One of the most vivid descriptions is of the Shekinah journeying with the people of Israel out of Egypt and protecting them from the pursuing Egyptians: a pillar of cloud by day and a pillar of fire by night. Exilic theology also made much good use of the doctrine, particularly in the concept of God's glory leaving the Temple and going to be with the people in exile. This highlights the capacity of the Shekinah to suffer with the people, to experience exile and rejection and then, in Ezekiel's closing vision, to return to the Temple, there to dwell forever in the midst of the people. Thus the 'Shekinah participates in the persecutions, and sufferings, and dying of the people, and together with that exiled people waits for its redeeming homecoming.'[70]

It is this vision of the Shekinah dwelling with the people of Israel forever, which inspires Moltmann's final vision of the consummation of all things. Creation itself becomes the dwelling place for God. The whole creation becomes the house of God. Whilst God and the world retain their distinctiveness, the doctrine of *perichoresis* means they also dwell within each other mutually.[71] They also participate in the attributes of the other.

69. Moltmann, *The Coming of God*, 302.
70. Moltmann, *The Coming of God*, 305.
71. This is Moltmann's adaptation of the concept of *perichoresis* – technically *perichoresis*

Creation participates in the divine nature of God, whilst God takes upon himself creation's limited time and space and gives creation eternal time and eternal space. In other words the presence of the cosmic Shekinah, dwelling within creation, makes all creation new. This new creation is then the new Jerusalem, God's dwelling place. God's unmediated and direct glory fills everything. All of creation is filled with the glory of God. God's glory becomes 'the inexhaustible source of creaturely life'.[72]

Perhaps the only question mark hanging over Moltmann's development of Shekinah theology is whether he really needs it. He suggests that the Christian contribution to Shekinah theology lies in the person of Christ, in whom we find a double indwelling: the indwelling of believers and the indwelling of God. Moltmann justifies the idea by suggesting that it can be expanded to say that 'creation is destined to be the dwelling space for God'.[73] Thus creation will be the dwelling space for God's glory (God's Shekinah) and the whole earth will be full of his glory. But it's not clear that Moltmann has established a need for a Christian theology of the Shekinah, as interesting as his developments are, when the doctrine of the Holy Spirit would seem to accomplish precisely what Moltmann wants to achieve in his Christian Shekinah theology. Moltmann indeed acknowledges the link with the doctrine of the Holy Spirit and suggests that Paul has combined the two doctrines in his letter to the Corinthians (1 Cor 6:19). The language Moltmann uses to describe the revealing of the Shekinah is also the language of the Holy Spirit, when he talks of 'The One' filling everything with his radiance.

4.9 Time

It is, however, the question of time, not space, which has brought the greatest criticism of Moltmann's eschatology. Moltmann has a great emphasis on being redeemed from the powers of history, and he includes evolution as one of the categories from which creation must be redeemed. As a result there is no place in his theology for salvation somehow emerging from the present. Salvation will come from the future, the coming God. Moltmann's entire eschatology is built on this premise. But his work depends on a particular understanding of the time of the eschaton and the future as the most influential and important category of time.

---

refers to a divine attribute.

72. Moltmann, *The Coming of God*, 319.

73. Moltmann, *The Coming of God*, 307.

The key concept upon which Moltmann bases his eschatology (and which is a contributing factor to his rejection of evolutionary Christology) is the idea of the end of time and the day of the Lord. He terms this the eschatological moment. This is the moment when the dead will all be raised at once, diachronically. Time will be no more; 'it will be gathered up, fulfilled and transformed through the eternity of the new creation'.[74] To describe this time, Moltmann resurrects the patristic and medieval term *aeon* (Greek) or *aevum* (Latin). This time is not the absolute eternity of God himself, however, it is the relative eternity of the new creation, which participates in the absolute eternity of God. Moltmann uses the same concept of God's indwelling in the time of creation as he does to describe God's indwelling in the space of creation. Creation then does not become timeless, but rather experiences the fullness of time, which is how we must understand eternity.[75] This creation as we now experience it is, obviously, a temporal one, which looks towards becoming an eternal creation. In fact its very temporality contains the promise of eternity, precisely because eternity is the fulfilment of time. The eschatological moment is the fulfilment of all created things, *including time*, which Augustine quite rightly saw as having begun with creation. Creation did not begin in time, but rather time began with creation. In that sense time was created by God and will be brought to fulfilment by God, along with everything else, which he created.

However, it is important to grasp the subtlety of Moltmann's model here, because saying that time began with creation is not the same as saying that God created time. If God created time, then 'does "the beginning" in which God created time fall in time or eternity?'[76] Moltmann insists that as long as time and eternity are defined as changeability and unchangeability (following Plato's model) this dilemma will continue. But he suggests that eternity need not be defined as simply the negation of temporality, but rather the fullness of creative life. Moltmann prefers not to describe this as God's divine self-determination to create the world, and, with the world, time. This self-determination would mark the 'beginning'. Rather Moltmann prefers the same model of events, which he also uses to describe God making space for creation: God's primordial self-restriction. God withdraws himself, making room for something outside to be created, something which is not eternal or omnipresent. The primordial mo-

74. Moltmann, *The Coming of God*, 280.
75. Moltmann, *The Coming of God*, 264.
76. Moltmann, *The Coming of God*, 281.

ment therefore is the moment when God chooses to restrict himself, and from the restriction of eternity emerges 'the time of creation'. This time fans out into future, present and past. This is an important question for eschatology, because Moltmann's concept of the eschatological moment depends on an understanding of the primordial moment. Understanding this helps answer some of the criticisms which have been made of Moltmann with respect to his concept of the eschatological moment.

Of particular interest is also the fact that the time of the created order is further divided into earthly and heavenly time. The course of time in the earthly creation is not cyclical or ending. It has an irreversible trend. As Moltmann mentions, this understanding of time has been explored further by current scientific discussions on both the nature of time and the second law of thermodynamics. The ideal of absolute time, which is itself timeless, simply doesn't exist in historical reality. Such time, which would be reversible, might be thought to exist in closed, symmetrical systems and systems in equilibrium: much like the vision of the future proposed by Ruether. But the second law of thermodynamics has introduced into physics the concept of irreversible time. Time does indeed march on and entropy is one result of the passing of time. Flows of energy are directional. Entropy, in fact, measures not just the loss of energy, but also the non-reversibility of the processes. Here one can see another basis for Moltmann's criticism of eco-feminist eschatologies, which seem to be based around cyclical notions of time. Time's arrow and the reality of entropy prevent any eschatology based on cyclical notions of time being ultimately convincing.

Moltmann does, however, acknowledge that earthly time is touched by heavenly time, which is indeed cyclical and eternal. Moltmann uses the term *aeonic* time here, to describe heavenly time as corresponding to the eternity of God, without actually being a part of God's eternity. Earthly time continually touches and is touched by the *aeonic* time of the invisible world. This earthly time looks with hope towards the future (so it is a time of promise), in which God will finally come to dwell within his creation and, in doing so, establish the eternal sabbath. The day of rest which we now celebrate is then a promise of that eternal sabbath, which will mark the consummation of all time, just as the arrival of the Shekinah will mark the consummation of all space. In this final consummation of time, the current transient time, which Moltmann sees as totally futile, will be transformed. The current temporality of creation, in which the power of ageing and death dominates everything, reflects God's absence from creation. Time passes and all that happens within time is 'irreversible, unrepeatable and inexorable'.[77]

In this context Moltmann suggests that the future, which he suggests is not the same as future *time*, must have pre-eminence among the modes of time. The future in this context is closely linked with the understanding of the eschatological moment and God's de-restriction of himself in his omnipresence and his eternity. Moltmann uses an ontological theory of time put forward by Georg Picht and Carl Friedrich von Weizsäcker, in which the modes of time have modalities of being. The 'past is the realm of reality, future is the realm of possibilities, and present is the interface at which possibilities are realised, neglected or suppressed'.[78] This ontological theory of time suggests that the future and the past are qualitatively different, that they are not homogenous in the sense of being able to be plotted along a single line. The understanding of future, which Moltmann's eschatology depends upon to a certain extent, is the 'transcendental condition for the possibility of time in general'.[79]

This notion of time as not being homogenous allows Moltmann to develop the notion of interruptions of time, particularly through recurring festivals, so that the Jewish Sabbath is understood as 'a messianic interruption of transitory working time' and the Christian Sunday 'as its eschatological interruption'.[80]

The key to understanding Moltmann here is his notion of the interlaced times, in which the present always means 'the relative simultaneity of past and future, through the force of remembrance and expectation'.[81] Eternity, as Moltmann defines it here, is the simultaneity of past and future, which is the aeonic eternity of the invisible world of heaven, and which reaches into time and interrupts time's flow. These eternal moments are moments of eternity in time, in which life is lived intensively and is totally fulfilled. They are moments of absolute presentness; all time becomes present and we experience the fullness of time. In such a moment of eternity the human being longs for a completely unclouded fullness of life, for the eternal present.

---

77. Moltmann, *The Coming of God*, 286.
78. Jürgen Moltmann, 'The Bible, the Exegete and the Theologian: Response to Richard Bauckham', in *God will be All in All: The Eschatology of Jürgen Moltmann*, edited by R Bauckham (Edinburgh: T&T Clark, 1999), 228.
79. Moltmann, 'The Bible, the Exegete and the Theologian', 228.
80. Moltmann, 'The Bible, the Exegete and the Theologian', 228.
81. Moltmann, *The Coming of God*, 290.

The ontology of time, which he develops, allows Moltmann to then develop his notion of the eschatological moment. Against Bultmann's notion of fulfilment and the meaning of history always lying in the present (where the present is experienced at the eschatological present by the Christian), Moltmann suggests that the present cannot be interpreted as 'the eschatological moment'. The eschatological moment no longer belongs to time at all. It is the opposite or, perhaps more accurately, the logical conclusion, of the primordial moment. It is the consummation of creation and the exit of time into eternity. So, just as Moltmann describes the primordial moment as God's restricting of his omnipresence and his eternity, so the eschatological moment is the de-restricting of these, that God might dwell in creation, and the creation might become eternal. All created beings will participate in God's eternity. *All* times will be transformed and transfigured and taken up in to the aeon of the new creation. Thus, both in terms of time and space, God's indwelling calls into being 'a kind of *perichoresis* of divine and cosmic attributes. In that new aeon a mutual perichoresis between eternity and time also comes into existence, so that on the one hand we can talk about "eternal time" and on the other about "eternity filled with time"'.[82]

Does the eschatological moment fall within time, and therefore history, or does it fall into the realm of eternity? Bauckham is not convinced that Moltmann has solved the dilemma of the delay of the parousia and the question of the timing of the eschatological moment. Bauckham suggests that the last day must, to a certain extent, 'also be the last date in the calendar, since it will bring all calendrical reckoning of time to an end'.[83] However, Bauckham's criticism depends to a certain extent on two lines in Moltmann's 1990 book, *The Way of Jesus Christ*, which do not appear in his later book, *The Coming of God*. Moltmann probably confuses the issue when he writes that the Day of the Lord 'is more than a day in the calendar' and 'will happen not only *in* time but also *to* time'.[84] Interestingly these particular phrases are absent from his later book, which otherwise shares many of the same arguments. In his later work Moltmann emphasises and develops the notion of the primordial moment as a way of understanding the eschatological moment (a theory he began to discuss in the earlier book) and omits the reference to the Lord's Day occurring in and to time.

82. Moltmann, *The Coming of God*, 295.
83. Richard Bauckham, 'Time and Eternity', in *God will be All in All: The Eschatology of Jürgen Moltmann*, edited by R Bauckham (Edinburgh: T&T Clark, 1999), 181.
84. Moltmann, *The Way of Jesus Christ*, 327.

The other problem which Moltmann doesn't address, which is related to the question of evolution and the processes of the world, is the value given to each time in history. If the eschatological moment simply gathers all time and transforms it, why do we bother to struggle along in our particular period of history? The solution seems too easy. God simply arrives in his eschatological moment, de-restricts his omnipresence and eternity, and transforms the whole of creation and time. It seems to devalue the integrity of each part of history. Scott Cowdell suggests that 'our world is a process as much of redemption as of creation, forever capitalising on opportunities, forever taking new directions after mishaps and restarting after calamities'.[85] God is at work in these processes, Cowdell suggests, 'always creating by constantly redeeming, in the physical, biological and human realms . . . '[86] Each moment of time, each process, has a part to play.

From an ecological point of view, the question must be asked of Moltmann's eschatology whether it values the integrity of every part of creation and time. This is a different question from asking whether every part of creation will be transformed. Certainly Moltmann is committed to the belief that the whole of creation will be transformed and that it will indeed be *this* creation, which is created anew. However, the key issue is, if every part of creation will be transformed, and every time will be transformed, do we really need to concern ourselves with environmental questions now? This is a particularly pertinent question to ask of Moltmann's work, because he suggests that 'the transformation into glory of this whole creation takes place diachronically, from the last day to the first, on the day of the Lord'.[87] But this concept, that everything will be transformed at once, in the eschatological moment, fails to value the integrity of the processes of creation. Can we really attribute any value to what we do, and to what all of creation does, if it will all be raised simultaneously? I suggest we cannot easily do so, which is why Moltmann's understanding of the eschatological moment doesn't provide us with an eschatological framework which is ecological.

4.10 Conclusion

An examination of Körtner and Moltmann, two key but different apocalypticists, suggests that there are serious flaws in apocalyptic eschatology when it comes to the question of ecology. Körtner's apocalypticism is not

85. Scott Cowdell, *A God for this World* (London and New York: Mowbray, 2000), 103.
86. Cowdell, *A God for this World*, 103.
87. Moltmann, *The Way of Jesus Christ*, 262.

hopeful enough, because any hope in the end is balanced by the possibility of total destruction. Whilst this approach lends itself to finding the balance of positive and negative apocalypticism, it leaves the reader wondering whether a fatalistic or nihilistic attitude to the rest of creation might indeed have some merit.

Moltmann's apocalyptic eschatology is ultimately more promising than Körtner's in fulfilling our desire to develop an ecological eschatology. The matter and time of this creation will be gathered up by God, created anew and will form the basis of the new creation. In this sense Moltmann's apocalyptic eschatology is promising, but the problem Moltmann faces is the question of continuity. How much can we talk of continuity, if this created order is completely re-created? Interestingly, though, Moltmann is not prepared to consider the possibility of total catastrophe, in the sense that Körtner is, but rather suggests the eschatological future will arrive in a series of particular presents, states or ages. These ages are transitions in the one eschatological movement of history. The temporal conditions of the whole of creation will be transformed. Thus the eschaton is the *consummation* of history, not its rupture.

His work on the times of transition potentially gives us an ecological model of eschatology, in the sense that, if we are in a time of transition, then everything that we undertake which is positive, both in human society and the environment around us, will contribute to a smooth transition to the kingdom of God. Most encouraging is his concept of *creatio continua*, which is an understanding of God the Spirit being present throughout the evolving world, encouraging self-transcendence, provoking new paths of evolution, through the Spirit's uncreated and creative energies. Moltmann's views on evolution are not consistent, however. He ultimately rejects the idea that the processes of self-transcendence or evolution could have a role in the final consummation of all things. This leaves his eschatology open to ecological criticism.

Moltmann's development of the Jewish concept of the Shekinah takes his eschatology in a much more ecologically friendly direction. The earth is valued as the bearer of the promise of something greater. Redemption comes from the mutual indwelling of God and the earth. We look to the earth as giving us a real experience of the promise to come and as being the locus of that promise. Creation itself becomes the dwelling place for God.

Addressing the problem of time, Moltmann further develops the medieval concept of the *Aeon* or *Aevum*. Time was created by God and will be brought to fulfilment by God, along with everything else, which he created. So, just as Moltmann describes the primordial moment as God's restricting of his omnipresence and his eternity, so the eschatological moment is the de-restricting of these, that God might dwell in creation, and the creation might become eternal. All created beings will participate in God's eternity. *All* times will be transformed and transfigured and taken up in to the aeon of the new creation.

However, Bauckham argues convincingly that Moltmann never really resolves the question of time. Moreover, his rejection of evolution as having any salvific qualities limits the importance he can place on creation, despite his protestations to the contrary. Any model of ecological eschatology must affirm, without reservation, the importance of the processes of the world, both human and non-human, in the final consummation of all things. Otherwise the question of relevance emerges. Moltmann comes very close to achieving this in his discussion of millenarian eschatology, but trips on the hurdle of evolution.

There is another model of eschatology which might provide a way forward. Moltmann hints at it when discussing the question of the resurrection of the dead. It is a concept which Karl Rahner explores when talking about the doctrine of intermediate state. He suggests that it is not a heresy to maintain that the perfection of human beings, body and soul, happens immediately after death and that the resurrection of the dead and the general judgement takes place parallel to the history of the world.[88] This concept has merit to it, particularly from an ecological perspective. To examine this theory of 'two worlds' I shall turn now to one of the key partners in the science and theology dialogue, John Polkinghorne, and to an emerging South African writer, Ernst Conradie.

88. Rahner, 'The Intermediate State', in *Theological Investigations* Vol XVII (London: Darton, Longman & Todd, 1981), 116.

Section III

Some models of the eschatological transformation of all things

Chapter Five
John Polkinghorne

5.1 Introduction

The question has been posed of Moltmann's theology, influenced as it is by apocalypticism (though Moltmann prefers the term 'apocalyptic eschatology' to describe his own position), whether, from an ecological point of view, it values the integrity of every part of creation, including all of time. Moltmann's theology goes a long way towards developing an ecological eschatology, but is limited by the fact that he doesn't view the processes of nature,[1] including evolution, as having any role in contributing to the final consummation of all things.

Nevertheless the ecological challenge, developed by Catherine Keller and Rosemary Radford Ruether, remains for contemporary eschatology. Any contemporary eschatological model must demonstrate that it passes the crucial litmus test of whether it is ecologically friendly, that is, it must address the issue of how the world is sustained and cared for. However, every potential model must not only be ecologically sensitive, it must also be theologically coherent. Moreover, if it is to be of any practical use in local churches and indeed world-wide communions of churches, it must show clear and demonstrable links with the Christian tradition. This last point is a crucial one. Academic theology needs to be at the cutting edge, as it were, but if it is to be usable by preachers or adult Christian education teachers, for example, it must at least be shown to have evolved from the tradition, rather than be something completely and radically new. This is a criticism that can be made of both Keller and Ruether, despite certainly Keller's protestations to the contrary.

1. The categories of 'nature' and 'creation' need to be distinguished. The first is a scientific category, the second a theological one. The theory of evolution properly belongs to the scientific category.

Developing an eschatological model that meets all these criteria is a relatively new exercise. Not a great deal has been written on ecological eschatology. One interesting model, however, that has been developed is the concept of two worlds, or multiple dimensions. This approach might also go some way towards addressing the inherent problems in Moltmann's work, that is, the question of the contribution human beings and creation itself make to the transformation of all things. John Polkinghorne and Ernst Conradie have both suggested that the transformation of all things can be understood as occurring alongside the world we see around us. This model allows us to comprehend what it might mean to talk of God's transcendence and our transformation, whilst not neglecting the fact that our human existence is rooted in the material (and therefore ecological) world. Critically, the new creation may not necessarily begin 'after' the ending of the old. 'The dead may be with God on one axis even while their dead bodies remain on earth on another axis,'[2] as Conradie puts it. For both Conradie and Polkinghorne this understanding of two worlds overcomes some of the difficulties surrounding concepts of time and eternity, as they relate to the eschatological consummation of all things.

5.2 John Polkinghorne

John Polkinghorne has developed his eschatology over a number of years now and has built on his earlier work with several more recent and substantial contributions.[3] He suggests that there are four propositions which undergird a viable approach to eschatological expectation.[4]

Firstly, and certainly fundamentally for his work, the universe, if it is indeed a creation, 'must make sense everlastingly, and so ultimately be redeemed from transience and decay'.[5] Here Polkinghorne develops one of his most helpful ecological contributions to eschatology, because the new creation arises 'as a redemptive act *ex vetere,* out of the old'.[6] Hence there is an element of continuity between the new and the old. Moreover, the

2. Ernst Conradie, 'Resurrection, Finitude and Ecology', in *Resurrection: Theological and Scientific Assessments,* edited by T Peters, R Russell and M Welker (Grand Rapids and Cambridge: Williams B Eerdmans, 2002), 289.
3. See especially John Polkinghorne, *The God of Hope and the End of the World* (New Haven and London: Yale University Press, 2002) and John Polkinghorne, *Science and the Trinity* (London: SPCK, 2004), especially chapter 6, 'Eschatological Exploration'.
4. See John Polkinghorne, *The God of Hope and the End of the World*, 148–149.
5. Polkinghorne, *The God of Hope and the End of the World*, 148.
6. Polkinghorne, *The God of Hope and the End of the World*, 116.

value of the old creation is clearly established because it 'affords the raw material for eschatological transformation into the new creation'.[7] Polkinghorne suggests that this emphasis on the new creation arising *ex vetere* prevents the kind of neglect of environmental duty threatened by a focus on other worldly hopes (a neglect which Keller and Ruether highlight). Whether the concept of the new creation arising out of the old is sufficient to allay Keller and Ruether's ecological concerns completely is something I discuss below.

Secondly, human beings must have a destiny beyond their deaths. Every generation must participate in that destiny. His development of the concept of God re-embodying the psychosomatic unity of the human person in the resurrection of the dead is helpful, particularly given the criticism of the idea of a bodiless, immortal soul rising to heaven. Moreover, he is clear that the human person does not exist in isolation, so that 'the "pattern that is me" cannot adequately be expressed without its having a collective dimension'.[8] By 'a collective dimension' Polkinghorne means the relationship between any human being and others, including the wider creation. He also acknowledges the need for a process of purgation and healing as a part of human beings' transformation.

Thirdly, we must acknowledge the tension between continuity and discontinuity, both on a personal level and a cosmic one. This is an important point to consider. The new creation, and the transformed human beings who are part of it, will not be totally new in the sense of sharing nothing in common with the old. So, for example, Polkinghorne suggests that the continuity between the old and the new creation will be such that the new creation will 'also have its "space" and "time" and "matter"'[9] but they will be everlasting. Yet the discontinuity means that the life to come will be free from the suffering and mortality that was such a significant part of the old creation.

Fourthly, the only basis of any hope for a destiny beyond death lies in the steadfast love and faithfulness of God. Here Polkinghorne rejects any concept of the new creation somehow evolving from the old, or of human beings bringing about their own eternity. Whilst the new creation arises from the old, it does not *evolve* from it, and the distinction is crucial in Polkinghorne's work. Hope depends on the action of God, not on the natural processes of the world. Thus he rejects any notion of creation itself

7. Polkinghorne, *The God of Hope and the End of the World*, 116.
8. Polkinghorne, *The God of Hope and the End of the World*, 109.
9. Polkinghorne, *The God of Hope and the End of the World*, 117.

participating in the transformation of the old into the new: any kind of evolutionary optimism is rejected. He therefore suggests that Teilhard de Chardin's picture of 'unfolding fulfilment with presence process'[10] is not without its problems, though he acknowledges that Teilhard de Chardin works from a credible theistic basis. Polkinghorne is not so charitable in his criticism of Tipler's[11] ideas, which are also rejected by him. Tipler suggests a physical eschatology, which could be achieved at the end of time by ever faster computers, resulting in an ultimate computer, which would resurrect human beings 'by creating computer emulations of them'.[12] But Polkinghorne suggests that such physical eschatology seems merely a ridiculous conclusion of evolutionary optimism. Rather we can hope in the faithfulness of God, who will not be thwarted, but will bring all things to fulfilment.

5.3 Background: the question of time

Polkinghorne's eschatology depends to a certain extent on his earlier development of the notion of the 'dipolar'[13] God. He develops this idea in order to articulate how God might be eternal and transcendent and yet intimately involved in creation and its time. God thus enjoys both eternity and temporality, that is, God is active in history and responsive to it, but not limited or constrained by his involvement in history. God is able, by nature of his dipolarity, to be both responsive to his evolving and suffering creation, as well as 'the guarantor of the order of creation and the ground of its hope'.[14] Polkinghorne uses the concept as a tool for understanding how God might interact in creation, particularly in the light of the changes in the understanding of time in the last century. There are, however, questions that need to be asked of Polkinghorne's understanding of transcendence, as I suggest below.

Polkinghorne reminds us that one of the great discoveries of the twentieth century was that the universe itself has a history; it is not timeless. A crucial theological question then became, how did God relate to the

10. Polkinghorne, *The God of Hope and the End of the World*, 25.
11. See Frank Tipler, *The Physics of Immortality: Modern Cosmology, God, and the Resurrection of the Dead* (New York: Doubleday, 1994).
12. Polkinghorne, *The God of Hope and the End of the World*, 24.
13. The concept is that God has two natures with respect to time: eternal and immanent.
14. John Polkinghorne, *Science and Providence: God's Interaction with the World* (London: SPCK, 1989), 80.

temporal flow of events, given that any divine action must take place in the developing processes of the world? In what Polkinghorne defines as the classical theistic doctrine of time, God observes all time in a simple act of perception. This model assumes that the entire history of the universe is laid out before God as some kind of four dimensional map, which he simply peruses. In that sense God's view of the history of the universe is eternal. The distinction between the past, present and future is therefore abolished. The world is a world of being, in this particular framework, 'but it is not a world of becoming. Nothing essentially new can ever happen within it'.[15] Whether Polkinghorne is entirely accurate in his description of the classical theistic doctrine of time is debatable. It does not necessarily follow that nothing new can happen under a divine eternal view.

Be that as it may, this is not the world which contemporary science describes, as Polkinghorne goes on. There is, as Polkinghorne describes the contemporary scientific view of the world, 'a delicate unpredictability of complex dynamical systems,'[16] which means that the future is not determined, and cannot be viewed as one might look at a map. The world is involved in a process of becoming, so that the future is indeed a part of our creation, and is not simply revealed to the traveller as he or she makes his or her way through life. The future's shape is responsive to our moulding. The consequence for God as described by classical theism is quite staggering. God does not know the future in the sense of being omniscient. Whilst God might be able to make highly informed conjectures, there are any number of eventualities which might emerge, for which he must be prepared. Polkinghorne suggests that, whilst God might have prepared plans for any eventuality, 'in his actual experience and knowledge he must be open to the consequences of the exercise of human free will and . . . the evolution of cosmic free process'.[17]

Polkinghorne suggests that the notion of the dipolar God prevents God from being in thrall to time. The notion of the dipolar God allows God to be involved in his creation, suffering with it, and yet also the provider of its hope. Polkinghorne suggests that this concept helps understand the dialectic of divine eternity and divine temporality. It prevents God from being either a remote tyrant, whose involvement within creation is minimal (the deistic understanding of God sometimes associated with Isaac Newton) or a pantheistic influence, a subjective God embedded in time.

15. Polkinghorne, *Science and Providence*, 78.
16. Polkinghorne, *Science and Providence*, 78.
17. Polkinghorne, *Science and Providence*, 79.

As I point out below, however, a non-contrastive[18] view of transcendence would also achieve this aim.

The dipolar view of God seems to contrast starkly with more established theological perspectives, particularly in trinitarian theology. Polkinghorne places himself at odds here with a key trinitarian conviction that the economic Trinity is the immanent Trinity. In recent theology this has been most fully developed by Karl Barth and Karl Rahner, as Duncan Reid has shown.[19] Barth was concerned to stress that the economic Trinity is the immanent Trinity. The doctrine of the Trinity is based solely on revelation. But God reveals himself as God is in himself. 'The trinity of God does not limit itself to the economy of revelation, because who God is in revelation, God is already in Godself (*zuvor in sich selber*). It is an outworking of God's absolute freedom that God does not reveal Godself other than God actually is.'[20] God therefore does not simply reveal information about God's self, but gives God's self. This is reflected in Barth's understanding that the three persons of the Trinity, or 'three modes of being' as Barth terms them, are indeed modes of *being*, not merely 'economic activities'.[21]

As Reid points out, this approach is pursued even more explicitly by Karl Rahner. Rahner is critical of 'the psychological theory of the Trinity', in which Christian theology has attempted to conceive of the inner life of the Trinity 'completely unrelated to us and to our Christian experience'.[22] He suggests that this psychological theory neglects our experience of the economy of salvation in favour of a 'seemingly almost Gnostic speculation about what goes on in the inner life of God'.[23] For Rahner this is unacceptable and contrary to the experience of revelation, which, he suggests, shows us who God is in himself. God's self-communication is exactly that, a communication of God as he really is. 'In the Trinity in the economy and history of salvation and revelation we have already experienced the

18. Kathryn Tanner, *God and Creation in Christian Theology: Tyranny or Empowerment?* (Oxford and New York: Basil Blackwell, 1988), 42. Tanner uses the term 'non-contrastive' in reference to one characterisation of divinity in Hellenistic theologies, but argues that it is applicable when discussing the apparent paradox in Christian theology of God's transcendence and creative agency.
19. Reid calls this concept 'the identity principle'. Duncan Reid, *Energies of the Spirit: Trinitarian Models in Eastern Orthodox and Western Theology* (Atlanta, Georgia: Scholars Press, 1997), 4.
20. Reid, *Energies of the Spirit*, 55.
21. Reid, *Energies of the Spirit*, 30.
22. Karl Rahner, *Foundations of Christian Faith: An Introduction to the Idea of Christianity* (New York: Crossroad, 1978), 135.
23. Rahner, *Foundations of Christian Faith*, 135.

immanent Trinity as it is in itself'.[24] God's self-communication to us is true self-communication. This leads Rahner to the accepted notion that the economic Trinity is the immanent Trinity and vice versa, or, 'the Trinity in the history of salvation and revelation *is* the "immanent" Trinity'.[25]

Polkinghorne is aware of some of the difficulties the notion of a dipolar God presents, because of the theory of special relativity. Most of us, it would be true to say, have an experience of time which is uniform. Yet there are in fact a multitude of different times, experienced by different observers relative to their states of motion. Moving clocks run more slowly than their inert counterparts. A clock which is shot off into space will return to earth to find its twin has grown older in the relative time elapsed. Unlike each individual, whose experience of time is necessarily limited to one, God experiences all time. Indeed Polkinghorne insists this is possibly the most important thing to say about God's relationship to the world, that he has a totality of experience. God experiences all time and every space. He has instant access to every event, wherever and whenever it occurs. Thus there is an absolute sense of God relating to time, despite the theory of special relativity. God's experience of time can in fact be understood as equating to cosmic time, which is the frame of reference used by cosmologists when speaking of the current age of the universe. In our expanding universe it is measured by the density of matter (averaged over many galaxies) or the temperature of the background radiation.

It would appear, though, that Polkinghorne's development of the notion of the dipolar God creates problems, which don't arise when a more traditional doctrine of God's transcendence is preferred. The issue of God's relationship to time, even time as described by contemporary physics, only becomes a problem for Polkinghorne because of his adoption of the notion of God's dipolarity. However, if one interprets transcendence in a non-contrastive way, immanence is not only implied but entailed.[26] There would then be no need for recourse to categories taken from Process thought. Polkinghorne seems to be assuming a contrastive understanding of God's transcendence, without allowing for a more comprehensive doctrine. Even if we accept, for example, the Orthodox position that there must be an absolute distinction made between the Creator and the creation, God's transcendence can encompass God's immanence. Or-

24. Rahner, *Foundations of Christian Faith*, 137.
25. Rahner, *Foundations of Christian Faith*, 136.
26. For a detailed discussion of this see Tanner, *God and Creation in Christian Theology*, especially chapter two 'The Transcendence and Creative Agency of God', 36–80.

thodox theology, building on the early church's development of the doctrine of *creatio ex nihilo* has repeatedly emphasised that there is an ontological distinction between the uncreated Creator and the created order. This distinction is reflected in the Orthodox doctrine of the divine essence (that which is totally unknowable) and the divine energies (God's revelation and reflection in the world).[27] Yet it does not necessarily follow that a contrastive understanding of God's transcendence is the inevitable result of the ontological separation of God from the world. The temptation 'to divide God from the world,'[28] as the Orthodox writer John Chryssavgis puts it, is not the only way forward. Traditional theism has a more embracing understanding of God's transcendence. Whilst emphasising God's transcendence, 'in that very inaccessibility God nevertheless remains immanent. In the radical otherness of God, the oneness of God in relation to the world is maintained.'[29] Thus, although the world remains a distinct reality from God, and God from the world, it is also 'a revelation of the personal and creative uniqueness of God. This revelation, at times silent and other times blatant, manifests an inviolable relationship with its Creator.'[30]

God's dipolar nature brings Polkinghorne to one of his most significant thoughts about God (not that it is unique to Polkinghorne). This is that God's perfection lies not in a notion of perpetual unchanging actuality, but rather in the idea of dynamical perfection. 'God's perfection is not the static holding of the topmost metaphysical peak, but it lies in the total love of his unfailing action.'[31] The rather beautiful example Polkinghorne uses is of a piece of music, which is not perfect when one particular moment is reached, as if we should attempt to hear that moment forever, but rather in the 'unfolding pattern of the whole movement'.[32] Within Godself, therefore, is a development. This notion of a developing or changing God, who is always perfect in his relating to the evolving cosmos, is yet another example of Polkinghorne's commitment to taking seriously the processes of the world. Another way of saying this is that dynamic perfection means that God continually surpasses Godself. God is surpassed by nothing other than God.

27. John Chryssavgis, 'The World as Sacrament: Insights into an Orthodox Worldview', in *Pacifica* 10.1 (February 1997): 13. See also Reid, *Energies of the Spirit*, 14–22.
28. Chryssavgis, 'The World as Sacrament', 13.
29. Chryssavgis, 'The World as Sacrament', 13.
30. Chryssavgis, 'The World as Sacrament', 15.
31. Polkinghorne, *Science and Providence: God's Interaction with the World*, 83.
32. Polkinghorne, *Science and Providence*, 83.

Polkinghorne is helpful in reminding us that the unredeemed universe is not everlasting. It has a history, a future, and an end. The notion of the everlasting nature of the universe is a problem, he suggests, for both those inclined to the notion of divine embodiment (some of the eco-feminists fall into this category) and also process theology. The point is well taken. In finding the balance between the integrity of the processes of the world and the vital nature of divine action in bringing all things to fulfilment, the essential contingency and indeed fragility of creation needs to be remembered.

Polkinghorne's concept of the dipolar God gives him a great sense of the unfolding processes of the universe. Consistent with his balancing of God's immanence and transcendence, Polkinghorne rejects the notion that God's involvement in time somehow deprives God of his sovereignty over creation. He suggests that 'what is beyond total control is not beyond total redemption.'[33] God's transcendence, which allows him his sovereignty over creation, is not lost in the notion of the dipolar God. However, again, it is Polkinghorne's contrastive notion of transcendence which creates this problem in the first place. A traditional theistic understanding of transcendence allows God to be both transcendent and immanent, without one unduly limiting the other.

In discussing how God relates to time, Polkinghorne also emphasises the significance of the doctrine of the incarnation. Using the analogy of a play, he suggests that God not only is the author *and* the producer (meaning that he has an ongoing relationship with the actors in the drama) but he has also appeared on the stage as one of the visible actors (in the person of Christ). The doctrine of the incarnation emphasises both Christ's humanity and his divinity. Jesus was fully human, and yet the Christian Church has always asserted that we encounter the saving action of God in Christ. This is the mystery of the incarnation: we encounter both the human and the divine in the person of Christ. God's kenotic action in Christ means, however, that although Christ was wholly divine, he was not the whole *of* the divine. The notion of the dipolar God allows us to comprehend each person of the Trinity possessing both his eternal and temporal pole. It was the temporal pole of the Son which became incarnate and gave up his eternal nature in the person of Jesus of Nazareth. Polkinghorne's argument here is somewhat problematic. It seems to move away from his

33. Polkinghorne, *Science and Providence*, 84.

dipolar notion of God. Moreover, it runs the risk of slipping closer to Arianism, even if this is not Polkinghorne's intention.

God's kenosis remains a theme in Polkinghorne's work for his understanding of divine action. He applies the concept of kenosis to God's omnipotence, eternity, omniscience, and causal status. Of interest to my work here is his suggestion that God has freely embraced the experience of time (the kenosis of eternity) and has restricted his knowledge (the kenosis of omniscience). The suggestion that God has freely embraced the experience of time corresponds to the notion of the dipolar God of eternity and time, but is explained from God's perspective by the kenosis of simple eternity. A temporal pole of divinity has been added to God's simple eternity, so that the Creator can truly engage with created time.[34] This kenosis applies also to God's omniscience, because the world is a world of true becoming; it is completely open to the future, which is brought about by 'intertwined causal principles, such as natural law, human agency, and special providence'.[35] By its very nature such a world is radically temporal. The future, in fact, does not yet exist, which means, logically, that even God does not know it.

Polkinghorne makes a distinction here between current omniscience, by which he means God knows everything there is currently to know, and absolute omniscience. This is an interesting point of debate, because it suggests that God does not see the as yet unformed future in any kind of intricate detail. Thus a limitation, in the form of the kenosis of omniscience, has been placed on God, but it is one that has been embraced from within the divine nature, not imposed upon God from without. In Polkinghorne's work the claim that God cannot foresee the future seems to be based on the idea that the future really is contingent. However, Polkinghorne seems to reject the traditional theological understanding that God's view of the future could embrace real contingency.

A criticism of Polkinghorne's notion of the dipolar God is that it results in undervaluing the very processes of creation, because God can take us, at death, from this world, and bring us across to his new creation. However, the theological basis of Polkinghorne's argument is convincing. He suggests that the self-restriction of God in the past in no way limits what God might do in the future. Whilst he rejects any notion of God being

34. See John Polkinghorne, 'Kenotic Creation and Divine Action', in *The Work of Love: Creation as Kenosis* edited by J Polkinghorne (London: SPCK and Grand Rapids, Michigan and Cambridge, UK: Eerdmans, 2001), 102–3.
35. Polkinghorne, 'Kenotic Creation and Divine Action', 103.

whimsically capricious about 'deciding to do something unprecedented, as if engaging in an act of divine showing off,' nevertheless he suggests, 'it is perfectly coherent to believe that in new circumstances ("the fullness of time") God will do new things'.[36] The miracle of the resurrection is the basis for the possibility of new acts of God. Indeed, Polkinghorne reminds us, already the divine act of the new creation has begun in the resurrection.

The inconsistency in Polkinghorne's writing, however, is that he is fully prepared to write about God's kenosis of causal status, which I discuss more fully below, but then is equally prepared to retract that idea when it comes to God's potential action in the future. This reflects the problems which result from his development of the idea of the dipolar God. We are presented with one doctrine of God in the present and quite a different one for the future. The God of the present has given up his causal status and become one cause among many, whereas the God who will bring about our future resurrection is still capable of genuinely novel and miraculous acts.[37]

It is interesting to note that Polkinghorne's later work on eschatology contains no explicit reference to the dipolarity of God. However, this understanding of God nevertheless informs his later work, particularly in the distinction he makes between the old and new creation, and the lack of an acceptance of God's sacramental presence in this 'old' creation.

5.4 The new creation 'ex vetere': evolution and cosmic process

Writing from the perspective of a scientist who later became a theologian, Polkinghorne is concerned to tackle some of the difficult questions arising from the theory of evolution, as they might pertain to eschatology. Polkinghorne suggests that any notion of evolutionary optimism, that is, that a satisfactory fulfilment will be found 'solely within the confines of the unfolding of present physical process',[38] must be challenged by the fact that the universe is condemned to ultimate futility, either in the form of a 'whimper' (cosmic decay) or a 'big crunch' (gravity takes total control). Moreover, every human death is simply the more immediate side of cosmic death. Thus, suggests Polkinghorne, as he gives his reason for examin-

36. Polkinghorne, 'Kenotic Creation and Divine Action', 105.
37. Polkinghorne, 'Kenotic Creation and Divine Action', 105.
38. John Polkinghorne, *Science and Christian Belief: Theological Reflections of a bottom-up thinker* (London: SPCK, 1994), 162.

ing eschatology, 'an ultimate hope will have to rest in an ultimate reality, that is to say, in the eternal God himself, and not in his creation'.[39] What Polkinghorne terms 'naturalistic evolutionary optimism'[40] is ultimately bankrupt, because 'the unfolding of present process alone will not bring about a golden age, for its history must peter out in final futility'.[41]

In contrast to the model of evolutionary optimism, Polkinghorne offers a fairly traditional eschatological model, but one which is worth exploring, because of his interest in science. He describes the soul as the 'infinitely complex, dynamic, information-bearing pattern'[42] of a person's being and life, which will be remembered by God as a psychosomatic unity and embodied by him in the transformed matter of this world. Consequently he affirms the soul's embodiment at the resurrection and suggests that this must occur in the transformed matter of the world. There is therefore a deep-seated intuitive hope which is cosmic in scope.

A question then arises for Polkinghorne, which is whether he allows for a sense of teleological purpose for this world. The promising title of one of his essays on the subject gives us some hope that he does.[43] Unfortunately the article itself focuses much more on the limitations of cosmic processes. Polkinghorne writes that there is a sense of futility about the cosmos as a whole. Not only are there more immediate dangers to the continuance of life on earth, such as the threat of a large meteor or a nearby supernova explosion sending lethal radiation onto the planet, as well as the realisation that the sun itself will turn into a red giant in five billion years, but the cosmos itself 'will end either in the dying whimper of decay or in the big bang of collapse in the melting pot of the big crunch'.[44] He puts the theological problem, which this reality engenders, quite succinctly: '... the knowledge that carbon-based life will eventually prove to have been but a transient episode in cosmic history puts a serious question to theology about the ultimate intentions of the Creator'.[45] Given his sense of futility in the cosmic processes and yet his belief in God's capacity to bring about a new creation, one question which Polkinghorne fortunately

39. Polkinghorne, *Science and Christian Belief*, 163.
40. Polkinghorne, *Science and the Trinity*, 146.
41. Polkinghorne, *Science and the Trinity*, 146.
42. Polkinghorne, *Science and Christian Belief*, 163.
43. John Polkinghorne, 'Eschatological Credibility: Emergent and Teleological Processes', in *Resurrection: Theological and Scientific Assessments*, edited by T Peters, R Russell and M Welker (Grand Rapids and Cambridge: Williams B Eerdmans, 2002), 43–55.
44. Polkinghorne, 'Eschatological Credibility', 47.
45. Polkinghorne, 'Eschatological Credibility', 47.

addresses is the issue of whether hope for a new creation devalues the present one. Why did God not simply start with the new creation? Polkinghorne's answer is quite conclusive, though, I will suggest, not necessarily helpful. He argues that the old creation is wholly other to God. God withdraws himself (*kenosis*) in order to allow the other to come into being and gives it its 'due independence to follow its own history'.[46] The old creation is therefore allowed 'to explore and realise its potentiality at some metaphysical distance from its Creator'.[47] But the new creation becomes a sacramental world, overflowing with the divine presence. It is 'brought into a freely embraced and intimate relationship with the life of God'.[48] The new is created from the old (*ex vetere*) and is the divine redemption and transformation of the old.

This model of the new creation coming from the old is potentially a very helpful one in terms of an ecological eschatology. Polkinghorne develops the idea that the old creation is free to grow and adapt. He suggests that the 'matter of this universe is perfectly adapted to its role of sustaining that evolutionary exploration of potentiality which is theologically to be understood as the old creation being allowed to "make itself"'.[49] He acknowledges that there is always a balance between order and disorder, and that 'new possibilities are realised "at the edge of chaos"'[50] with the cost of novelty being 'the certainty of its impermanence'.[51] However, it is indeed this evolving creation which is then transformed into the new creation. The concept implies that the evolution of this creation is significant in an eternal sense, that is, that what happens in this creation matters eternally.

Unfortunately the way Polkinghorne further develops this idea is extremely problematic. In stressing the metaphysical distance of the Creator from the creation, Polkinghorne is implying that the old creation lacks God's presence. It effectively limits God's immanence and any notion that the old creation experiences the sacramental presence of God in the world, because the emphasis is on the new creation being a sacramental world. This is illustrated by his comment, 'I do not accept panentheism

46. Polkinghorne, *Science and Christian Belief*, 167.
47. Polkinghorne, *The God of Hope and the End of the World*, 116.
48. Polkinghorne, *The God of Hope and the End of the World*, 116.
49. John Polkinghorne, 'Eschatology: Some Questions and Some Insights from Science', in *The End of the World and the Ends of God: Science and Theology on Eschatology*, edited by J Polkinghorne and M Welker (Harrisburg, Pennsylvania: Trinity Press, 2000), 39.
50. Polkinghorne, 'Eschatology: Some Questions and Some Insights from Science', 39.
51. Polkinghorne, 'Eschatology: Some Questions and Some Insights from Science', 39.

as a present reality, but I believe it will become an eschatological reality'.[52] But herein lies a significant ecological problem for Polkinghorne. It is precisely this kind of attitude, often wrongly attributed to Christianity as a whole, which views the natural world as completely lacking the presence of God. This results in the natural world being viewed as something to be exploited, as eco-feminist theologians, amongst others, have reminded us. It denies an important theme of ecological theology: God's creative and life-giving presence to all creatures. It is not typical of Anglican thought. Given that Polkinghorne is an Anglican priest, it is a curious addition to his theology. As I will suggest in chapter eight, there is much to be gained for eschatology by adopting an explicitly sacramental view of this 'old' creation. The Eucharist is one example of the sacramental nature of this world.

The tradition, suggests Polkinghorne, gives us clues as to what this new creation, which is created *ex vetere*, will mean. The incarnate, cosmic Christ is the path for God's universal return. The cosmic Christ brings the old creation into this relationship in the life of God. Notions such as panentheism, and the Eastern Orthodox idea of the whole destiny of creation being *theosis* (deification), are also helpful in understanding what God's sacramental presence in the new creation will mean. Typically, though, the word panentheism applies to this world rather than the next. Polkinghorne suggests that there is genuine independence in the old creation, but that ultimately God will bring about the new creation. This concept of the new and old creations, he suggests, is an attempt to do justice to the God of process, working in the old creation, and to the God of hope, who brings about the new creation. This God of hope is the ultimate answer to present suffering, because 'each generation must receive the healing and fulfilment that is its due'.[53] Polkinghorne's language is uncompromising at this point. To suggest that every generation 'must' receive healing is a much more forceful statement than we 'hope' it will be so.

The concept of the new creation coming from the old (*ex vetere*) is not necessarily a helpful one for the purposes of developing an ecological eschatology. Whilst it does indeed affirm the value of the old creation, as Polkinghorne suggests, whether it is helpful to describe the old creation as the 'raw material'[54] of the new is debatable. Whilst the term implies that the 'stuff' of creation and the creatures themselves are important, it also

52. Polkinghorne, 'Eschatology: Some Questions and Some Insights from Science', 40.
53. Polkinghorne, *Science and Christian Belief*, 169.
54. Polkinghorne, *The God of Hope and the End of the World*, 116.

carries a negative connotation about the processes of the old creation. The old creation is seen exploring its potentiality at some distance from the Creator.[55] But this very description of creation's processes and activities implies that they simply help creation's own learning or development, but have no eternal *contribution* to make. In Polkinghorne's theology neither creation's contribution, nor human beings' contribution as a part of creation, are operative in an eschatological sense.

5.5 Destiny beyond human beings' deaths: personal and 'typical' destiny

In a very helpful way Polkinghorne does articulate a way of comprehending God's resurrection of human beings. Influenced in his work on God's memory to some extent by process theology, he suggests that every part of our lives is significant to God, including all of the relationships which we form, as well as our physical bodies. All of this forms a pattern of information, which is remembered perfectly by God, who will 'then reembody it in the ultimate divine eschatological act of resurrection at the last day'.[56] He does not explicitly extend the model to all of creation, which is what I believe must occur. However, his understanding of God re-embodying the pattern that is every human being is a good one. Importantly in Polkinghorne's eschatological model, God does not override human freedom. Some may still choose hell, where the doors are locked from the inside! But, for those who choose, everlasting life awaits.

In Polkinghorne's understanding of our heavenly destiny lying somehow parallel to the world we see around us, we must all be purified in some way, before we come to our eternal destiny. This purgation 'must be part of the transforming process that fits human beings for everlasting encounter with the reality of God. It will surely not be brought about by an instantaneous act of divine magic.'[57] Nevertheless 'the sinful dross of our lives must be refined away'.[58] As dramatic as that language is, Polkinghorne is thinking of God's divine forgiveness, steadfast love and mercy here, rather than of more fiery imagery such as the wrathful potentate. Thus God's purgation of us will be like the work of a healing surgeon.

55. Polkinghorne, *The God of Hope and the End of the World*, 116.
56. Polkinghorne, *The God of Hope and the End of the World*, 52.
57. Polkinghorne, 'Eschatology: Some Questions and Some Insights from Science', 41.
58. Polkinghorne, *Science & Christian Belief*, 170.

God will, however, not allow anything good to be lost and he expects a 'personal destiny for each one of us',[59] whilst expecting a more 'typical' destiny for the cosmos and other created beings. Yet Polkinghorne affirms a general eschatological purpose for all of creation. He does not elaborate on his ideas here, which is a pity, because his work remains anthropocentric. A more satisfactory approach, to my mind, would have been to affirm the inter-connectedness of all living things, and to leave it at that. Instead he seems to try to guess what particular approach God will take to individuals and the rest of his creation. So we are left with the rather curious idea that, whereas human beings will enjoy an individual destiny, with other creatures it is more the idea of the species being preserved which seems to be acceptable: 'With the animals we attached significance rather more to the species than to the individual. We would be sorry if all deer were exterminated, but the culling of individuals to preserve the herd seems to most of us to be an ethically accepted practice.'[60]

Here then the limits of Polkinghorne's theology begin to be exposed. The problem with his theology here may be demonstrated in a brief reflection on the current debate on the Japanese Fishing Authority's proposal to resume commercial whaling. This debate demonstrates the different ethical perspectives being employed by each side. From the pro-whaling perspective, the argument is that as long as human beings don't threaten the various whale species, in the way that they did until the cessation of commercial whaling in the late twentieth century, then there is no problem. Individual whales don't actually matter. The ethic being espoused here is one of survival of the species, but not of the individual value of any particular animal. Opposed to this are the anti-whaling nations, whose ethical basis is that each individual whale is an intelligent creature, with whom we are linked by our interconnectedness as intelligent mammals. The life of each whale, therefore, is valuable in itself. But this approach is also flawed, because one could just as easily make the same argument about domestic cattle and sheep. Having said that, though, human beings *do* make distinctions between various creatures; we value individual pets, as well as such creatures as dolphins and whales, more than other creatures such as cattle and sheep.

The above discussion highlights the inherent complexities in discussing the issue of the eternal destiny of particular creatures. The danger with Polkinghorne's theology here is that it is excessively anthropocentric. It

59. Polkinghorne, *Science & Christian Belief*, 172.
60. Polkinghorne, *Science & Christian Belief*, 172.

is precisely this approach, however, which has caused so many environmental problems. The downplaying of the connection between all living things easily leads to the attitude that the natural world can be exploited. Interestingly enough Polkinghorne's thinking develops in a minor way on this particular issue, with his more recent essay displaying something of an acknowledgment of the issue posed by his earlier work. In his later essay he suggests that theology has often been excessively anthropocentric in its thinking and that a better way to imagine the fulfilment of all things is that 'all creation must matter to the Creator in ways that are appropriate to its nature. Therefore all creatures must find their true fulfilment.'[61] Whilst every bacterium that has ever lived is not seen as being resurrected to an everlasting destiny, he does acknowledge that neither is it only human beings which matter to God.[62]

Thus, on the positive side, Polkinghorne affirms that there is hope for all. There is a destiny for all matter as well as human beings. All of creation is important to God. He suggests that if one's eschatological vision is not cosmic in scope, then those non-human parts of creation would appear to have no lasting value, and 'ecological concern would then be seen simply as a matter of prudent human provision for its own needs'.[63] It's an interesting ethical point, whether creation has any intrinsic and lasting value, if one has no eschatological vision. Certainly, as Keller and Ruether have reminded us, if the focus of eschatology is on human beings alone (their criticism of traditional eschatology) then creation is available for unlimited exploitation. But I suspect Polkinghorne is also correct, when he asserts that creation would seem to have no lasting value if one's eschatology is not cosmic in scope.

5.6 Continuity and discontinuity: The 'intersection' of the historical and eschatological

As I have shown above, Polkinghorne has engaged in fascinating discussions about time in his work on eschatology and one of the questions he poses is whether the new creation comes about after the old or whether it is parallel to it and contemporaneous with it. In traditional theology

61. Polkinghorne, 'Eschatological Credibility: Emergent and Teleological Process', 49.
62. For a more complete and interesting discussion of the destiny of all creatures, see Denis Edwards, 'Every Sparrow that Falls to the Ground: The Cost of Evolution and the Christ-Event', in *Ecotheology* 11.1 (March 2006): 103-123, especially 119.
63. Polkinghorne, 'Eschatological Credibility: Emergent and Teleological Process', 49.

it occurs after the old. The remembered patterns of our souls are held in the mind of God until the resurrection and transformation of the entire cosmos. In process theology our entire future is indeed perceived in terms of being remembered by God. But Polkinghorne suggests, correctly in my opinion, that traditional theology offers much more than that. We will rather experience the beatific vision. And, 'in the unveiled openness of the new creation to its Creator',[64] we will, in perfect freedom, embrace and be embraced by the divine will. This is a much richer understanding of the promise of eternal life. Moreover, he comments that the concept, that the new creation necessarily follows on consecutively from the old, suggests that the two worlds need to share a single time. Polkinghorne argues that 'their mutual relationship can be more subtle in its character than that'.[65] It may be possible to conceive of the two worlds existing parallel to each other. It is to this significant idea that I now turn.

Polkinghorne's exploration of the relationship between the old creation and the new creation, or the historical and the eschatological, is perhaps his most refreshing work in the field of eschatology. He uses a mathematical model of eschatological fulfilment when he writes of 'some degree of *intersection* between the historical and the eschatological'[66] [my italics].Using mathematical imagery, he suggests that the old and new creations could be thought of as 'vector spaces orthogonal to each other (the old and new creations) but connected by mappings or projections (resurrection, redemptive transformation) or even intersections (resurrection appearances)'.[67] Such intersections may be perceived, suggests Polkinghorne, in the resurrection appearances and even in the Eucharist. Indeed, through the resurrection Jesus Christ has become 'the unique link between the life of God and the life of creatures, the bridge by which entry into the divine life becomes possible for us, so that it is in Christ that eschatological fulfilment will be attained'.[68] Whilst this concept may seem somewhat alien at first to the theologically trained, Polkinghorne suggests that for those with a mathematical background it would seem to be as helpful an image as any. Polkinghorne's comment, that we time-bound creatures are necessarily limited in our thinking about how multiple dimensions of time might relate to each other, is well taken.

64. Polkinghorne, *Science and Christian Belief*, 174.
65. Polkinghorne, 'Eschatological Credibility: Emergent and Teleological Processes', 53.
66. Polkinghorne, *Science and Christian Belief*, 170.
67. Polkinghorne, *Science and Christian Belief*, 173.
68. Polkinghorne, *Science and the Trinity*, 166.

Yet, whilst I found Polkinghorne's argument about orthogonal vector spaces as a model for understanding the 'time' of eschatology fascinating, particularly the idea that the resurrection appearances are examples of the intersection between the historical and the eschatological, ultimately the model has more problems than strengths. There are several criticisms that can be made of Polkinghorne's work. Firstly, the model of multi-dimensions seems unnecessarily complex and speculative. Whilst it must be accepted that any eschatological model is by its very nature highly speculative, nevertheless to have two dimensions or worlds existing side by side seems to introduce yet another factor into our belief system; the model introduces extraneous material, as it were.

Another significant problem potentially exists with his concept. If we accept this model, which speculates how human beings and creation enter into God's eternal life, what does this say about the value of each particular moment within history, and each particular life's contribution? Like many theologians, including Moltmann, Polkinghorne struggles to maintain the balance between affirming the value and significance of the processes of the world, and affirming God's omnipotent power to save. Polkinghorne's dilemma in this respect can be seen in his reluctance to spell out too clearly how the new creation is related to the old. Polkinghorne quite neatly summarises the dilemma which faces every work on eschatology:

> It may be that the time of that new world bears no relation to the time of this world and that we pass to experience of it directly at death. Or it may be that that time will only begin to run when this world's process will have been concluded by some great act of God, so that the new world comes into being by the transformation of the old.[69]

In this context, the sacraments, particularly the Eucharist for Polkinghorne, provide us with a foretaste of that final transformation. This foretaste reveals the divine nature of God and Christ, who will eventually be 'all in all'.

69. Polkinghorne, *Science and Providence*, 90.

5.7 Hope lies in the steadfast love and faithfulness of God

In his book, *Science and Providence,* Polkinghorne works hard to find the balance between the eternal future of creation, God, and the relevance and importance of temporal creation and its processes. In his chapter on 'hope', he identifies the two main possible futures of the universe (according to scientific prediction) as being bound up in a continuing tug-of-war between the expansion caused by the explosion of the big bang, and the contractive effect of the force of gravity. Acknowledging that we really don't know which force will triumph, he suggests that neither prospect in itself looks particularly promising. Nevertheless, using the starting point that ultimate hope must be grounded in God alone, he suggests that 'hope must also include hope within the process of the world itself. The fulfilment of God's intentions cannot wait only on a final act by which all things are made new.'[70] The criticism he makes of eschatological hope, which emphasises a final act of God, is that it is always in danger of becoming 'pie in the sky'. This is consonant with the criticisms made by Keller and Ruether. It turns God into a god of magic and contradicts our knowledge of God as working patiently in creation, though it has to be pointed out here that all such claims about God working in creation are informed by faith. 'The clue to the nature of our *present* hope lies exactly in God's continuing interaction with the world that he has made.'[71] In agreeing with Arthur Peacocke's concept of God being 'an Improviser of unsurpassed ingenuity'[72] he suggests that hope does not lie in God's revealing an inexorable plan, which God has calculated and ordained in every detail, but rather 'in the evolving potentiality of a universe in which chance and necessity maintain their fruitful interplay and within whose open process the immanent activity of God is at work'.[73]

In this work Polkinghorne maintains a good balance between, on the one hand, an acknowledgement of the importance of the continuing existence of the world and its processes, and, on the other, God's ultimate fulfilment of his purposes in creation. Thus, drawing on his development of the dipolar God, he suggests that the 'infinite God is a source of inexhaustible hope, both within the processes of this world and beyond it'.[74] God does not know the future, in the sense of being absolutely able to

70. Polkinghorne, *Science and Providence*, 97.
71. Polkinghorne, *Science and Providence*, 97.
72. Arthur Peacocke, *God and the new biology* (London: Dent, 1986), 98.
73. Polkinghorne, *Science and Providence*, 97.
74. Polkinghorne, *Science and Providence*, 98.

control it, otherwise the processes of the world would have no integrity of their own. We would be living in a controlled world, whose future was already determined. God *is* affected by his creatures' actions, in the sense that he must respond to them.

It's important to point out here again that Polkinghorne's theological understanding does not include the concept that God's view of the future could embrace real contingency. Knowing the future is not the same as having absolute control over it. However, I agree with his conclusion that, ultimately, God will not be thwarted, but will bring all things to fulfilment.

Nevertheless Polkinghorne's theology is limited here, because he makes no provision for the processes of the world to participate in any way in their eternal inheritance. As I show below, he makes a distinction between the contribution human beings and the rest of creation make now, in this life, and the bringing about of our eternal inheritance

One helpful phrase Polkinghorne uses here is 'divine resourcefulness'. Nothing is beyond God's ability. The future, whilst not predestined in an absolute sense, is nevertheless assured. God's purposes will be fulfilled, one way or another. The strength of Polkinghorne's work here is the balance he finds between the freedom of creation, the value of the ongoing processes of creation, and God's future. God is leading all of creation into his future.

Yet, despite his considerable work on this particular issue, the problem still affecting Polkinghorne's concept of our transference to a new creation (wherever this might lie in time) is the devaluing of *this* creation that occurs as a result. We see an incompleteness in Polkinghorne's work here, because in fact he *does* see a place for God's creatures in bringing about the future.[75] Yet this appears to be limited to the sense of *temporal* future. In that respect Polkinghorne seems to follow a similar path to Jürgen Moltmann. Moreover, Polkinghorne allows for 'an interweaving of providential and creaturely causalities',[76] in order to bring about an intrinsically unpredictable and therefore open future. God is, in the history of an evolving creation, simply one cause amongst many. In other words, God is not always in total control. The self-limitation of God's nature is manifested perhaps most clearly in the kenosis of causal status. The Incarnation is the most striking example of this, in which we see God submitting 'in

75. Polkinghorne, *Science and Providence*, 98.
76. Polkinghorne, *Science and Providence*, 98.

the most drastic way to becoming a cause amongst causes'.[77] God is willing to share with creatures in a profoundly self-limiting way.

Polkinghorne deliberately abandons the traditional distinction between primary and secondary causality, because he suggests that classical theology does not adequately reflect God's willingness 'to share with creatures',[78] in the sense of allowing them 'their part in bringing about the future'.[79] There are, however, negative implications, which arise from the suggestion that God is one cause amongst many, which I discuss below.

Part of the cause of the problem for Polkinghorne seems to arise because of his desire to establish a 'causal joint' between the Creator and the creation. Polkinghorne is reacting particularly to Thomist thinkers, for whom, he suggests, it is enough simply to state the theory of primary and secondary causality, without attempting to explain how divine causality occurs in and under the creaturely causalities. Out of a desire to avoid theology and science using two different languages and talking 'past each other at different levels',[80] Polkinghorne strives to describe how we might understand this causal joint. One of his attempts has been to describe God's action as the working of 'pure Spirit',[81] in the sense of '*pure* information input'.[82] He suggests this was precisely in an effort to avoid the criticism that he had reduced God's role to being simply one cause amongst many.

On his own admission, though, he has more recently pursued a slightly different line of thinking, which deliberately describes God's role as indeed one cause amongst many, based on his understanding of God's 'kenosis of causal status'.[83] He suggests that this is God's loving choice, by which divine and creaturely causalities become interwoven. Polkinghorne also suggests that this makes it possible 'to conceive that this kenotic providential causality is also exercised energetically as well as informationally'.[84] Curiously, despite all his work on the kenosis of divine causal status, he still prefers the concept of the working of the Spirit being identified with pure information input.[85]

77. Polkinghorne, *Science and Providence*, 104.
78. Polkinghorne, 'Kenotic Creation and Divine Action', 104.
79. Polkinghorne, 'Kenotic Creation and Divine Action', 105.
80. Polkinghorne, 'Kenotic Creation and Divine Action', 97.
81. Polkinghorne, 'Kenotic Creation and Divine Action', 101.
82. Polkinghorne, 'Kenotic Creation and Divine Action', 101.
83. Polkinghorne, 'Kenotic Creation and Divine Action', 104.
84. Polkinghorne, 'Kenotic Creation and Divine Action', 105.
85. Perhaps the mathematical physicist still has to surface now and again!

Polkinghorne's emphasis on the kenosis of the divine causal status makes God a secondary cause, one amongst many. Yet in doing so he produces the unintended consequence of potentially reducing God to a pantheistic being, who is only able to interact within creation, and even then in an extremely limited way. The problem of the nature of God's being emerges. Moreover, it raises questions about Polkinghorne's eschatology, which is precisely that hope lies in the steadfast love and faithfulness of God. What hope is there, ultimately, if God's providential action is limited to being one secondary cause amongst many? Yet Polkinghorne *wants* to affirm that, ultimately, hope comes from God and God alone. So we see the internal contradictions which emerge in his work. On the one hand God gives up his causal status within creation, and, on the other, hope ultimately lies in the steadfast love and faithfulness of God. But this then suggests that there are two natures of God, which is exactly what Polkinghorne explores in his work on the dipolar God. I have already discussed some of the problems with that concept.

There are other approaches to the question of divine causality, which avoid the problems Polkinghorne faces. A more embracing understanding of traditional theism would permit us to comprehend how God might be the primary cause of all things, and yet still be intimately involved within creation. Scott Cowdell has further developed this approach, using the traditional understanding of primary and secondary causality, but within the context of a panentheistic model, with provision for special providence. The doctrine is therefore that of double agency. Within this context God's special acts 'are constrained by the nature of the world which God is creating', which goes some way towards what Polkinghorne is suggesting in his exploration of God's kenosis. Yet Cowdell allows for the notion of God's general providence, the primary cause of everything, understood in terms of the contemporary scientific description of the world as 'the general providence at work through a process of complexification and adaptation'.[86]

From an ecological point of view, another problem with Polkinghorne's theology is that his concept of God's kenosis is not translated into creatures playing a part in their *eternal* inheritance. This appears in his theology to be all *God's* work. Polkinghorne is right to suggest that an eschatological hope cannot 'be based on a kind of evolutionary optimism that depends *solely* on the extrapolation of the present process'[87] [my italics].

86. Scott Cowdell, *A God for this World* (London and New York: Mowbray, 2000), 113.
87. Polkinghorne, 'Eschatological Credibility: Emergent and Teleological Processes', 48.

But unfortunately his eschatology tends towards the other extreme, and depends completely on God's faithfulness '*alone* [being the] ground for an everlasting hope'[88] [my italics]. This element of Polkinghorne's theology is all the more curious, because his work on the elements of continuity between the old and the new creation and his understanding of the evolutionary character of creation points him in a different direction. Indeed he asserts that the gentle and developmental nature of the old creation is precisely the form that our future eschatological fulfilment in heaven will take: 'In fact, there can surely be no other way in which finite creatures can adequately encounter the infinite reality of their Creator in a non-destructive fashion.'[89] But, to place these comments in context, here he is referring to human beings getting to know God (to put it succinctly) in the after-life. The life to come 'will not be a timeless moment of illumination, but it will be the everlasting exploration of the inexhaustible riches of the divine nature.'[90] Yet Polkinghorne never takes the next logical step, which is to allow for the possibility of the processes of creation playing *some* part in the eternal destiny of creation. Certainly the model of the two worlds, old and new, having a more subtle relationship with each other, than one simply following consecutively upon the other, allows for this gentle and developmental nature of the new creation. However, it is a concept which Polkinghorne doesn't pursue.

5.8 Conclusion

Polkinghorne contributes a number of ideas to the development of an ecological eschatology. His focus on the universe having an eternal destiny, whilst being aware of the difficulties raised by scientific predictions about the ultimate futility of the universe, is a helpful one. He allows a scientific 'bottom-up' contribution to eschatology, which is crucial given the focus of Keller and Ruether on the negative aspects of eschatology as it pertains to ecology, and their assertion that traditional eschatology has focussed too much on other-worldly matters. Polkinghorne's work gives an air of scientific credibility, or possibility, to an area of theology which is still necessarily speculative. His emphasis on God's kenosis, which allows creation to explore its own potentiality, to 'make itself', gives due emphasis and importance to the freedom of creation and the significance of its

88. Polkinghorne, 'Eschatological Credibility: Emergent and Teleological Processes', 48.
89. Polkinghorne, 'Eschatological Credibility: Emergent and Teleological Processes', 55.
90. Polkinghorne, 'Eschatological Credibility: Emergent and Teleological Processes', 54.

processes. His commitment to ecology means that he is concerned to explore eschatological criteria which emphasise the continuity between the old and new creations, as well as the discontinuity. His work also reminds us of some of the shortcomings and failings of the old creation and its need for redemption. This includes the purgation of human beings, that they might be re-embodied in the final resurrection. Moreover, his emphasis on God's steadfast love and faithfulness is a helpful reminder that eschatology depends on God's action, and that we ought not to place too heavy an emphasis on the natural processes of the world bringing about an eschatological transformation. His work on time and the possibility of the new creation taking shape parallel to the old, and yet arising from the old (*ex vetere*), is helpful too, because it allows for the concept of the new creation gradually emerging from the old.

There are a number of weaknesses with Polkinghorne's eschatology, which in the end limit his contribution to an ecological eschatology. His understanding of God's transcendence is too limited, and results in his exploration of the highly problematic understanding of the dipolar nature of God. Whilst his work on God's kenosis has the positive effective of giving due weight to the natural processes of the world, his work on God's kenosis of his causal status reduces God to being one cause amongst many. This is a further limitation of God's transcendence.

Ultimately Polkinghorne's contribution to eschatology is not satisfactory from an ecological point of view, because he never articulates *how* the processes and history of the world might be integrated into the new creation. His focus seems to disallow the processes of creation having a role to play in the consummation of all things. He does suggest that God's total creative intent seems to be a two step process, with the new creation arising *ex vetere*. However, his suggestion that the old creation is described as being allowed to explore its potentiality at some 'metaphysical distance from its Creator'[91] has difficulties. It seems to deny the possibility of God's sacramental presence within his creation. In his schema it is only the *redeemed* creation which is brought into an intimate relationship with the life of God. The question also needs to be asked of Polkinghorne whether the processes of creation have a place in that redemption. If the new creation does indeed take shape alongside the old one, rather than at the end or the completion of the old one, then we need to state *how* this creation and its history and processes are valued. Surely there is a place

91. Polkinghorne, *The God of Hope and the End of the World*, 116.

for the processes of creation *contributing* to the gradual passing of the old into the new. There is one writer who has explored some possibilities in exactly this direction and so I now come to an examination of the work of the South African writer, Ernst Conradie.

Chapter Six
Ernst Conradie

6.1 Introduction

Before I come to those elements of Ernst Conradie's theology, which help me develop further some of Polkinghorne's ideas on the 'two-worlds' theory, it is important to set Conradie's work in context. Conradie acknowledges how much his own background inevitably influences his writing. He writes as a 'white, middle-class, male, Africaans speaking, reformed South African'.[1] Conradie's early life was relatively sheltered, it would seem. He grew up enjoying the exquisite natural beauty of the Jonkershoek valley near Cape Town. His childhood experience helped him appreciate such beauty and his training as a minister in the reformed tradition guided him to see this beauty as God's creation. However, his wider experience helped him realise that the rest of the world is not always as beautiful as the valley of his childhood. He was exposed to the stark contrast between that beauty and the South African realities of poverty, apartheid and naked racism and oppression. His contact with 'ugliness' around him indubitably influences his writing, as he acknowledges. It is a timely reminder for those of us perhaps less exposed to the ugliness of the world and the reality of sin. Any model of eschatology which seeks to describe, in an ecologically sound way, the transformation of all things into the kingdom of God, must come to terms not only with the finitude of the creation, but its deep seated sin as well. As someone brought up in the culture and gentleness of Adelaide, and now living in the comfortable, sometimes sleepy (in the winter) town of Albany on the southern coast of Western Australia, it is a perspective one needs to remember and integrate into one's eschatology.

1. Ernst Conradie, *Hope for the earth: Vistas on a new century* (Bellville, South Africa: University of the Western Cape, 2000), 11.

In fact there are three human predicaments to which Conradie sees eschatology traditionally responding: firstly, the predicament of human self-enclosure due to the effects of sin in various manifestations of evil; secondly, the predicament of finitude in time: mortality and transitoriness; and thirdly, the predicament of finitude in space, which has important implications for the limitations of human power and knowledge.

One temptation for all theologians, which arguably Ruether and Keller in fact fall victim to, is to view the world as being inherently more positive than it is. Not that either of those authors is unaware of the effects of human sin, but perhaps they do not take sufficient account of the other predicaments which face us, including the finitude of creation itself. Conradie's work is valuable in reminding us of those predicaments and entering into a detailed examination of them in his eschatology, rather than glossing over those particular issues.

6.2 The predicament of evil: the restrictive effects of sin

Unlike the predicaments of temporal and spatial finitude, the predicament of evil is entirely due to the self-infliction of human self-enclosure. The effect of sin, through the many and varied manifestations of evil, are destructive, to say the least. Ecological devastation is one such effect. For the most part evil manifests itself in societal structures that are created by the powerful to subdue the powerless and are used to 'enforce domination, exploitation and alienation'.[2] Sometimes the processes of violence and oppression are indeed legitimised through the various structures and systems of society. Of particular interest to this book is the effect of sin in the form of material greed, which can have such a devastating effect on both the environment itself and on other human beings.

Conradie highlights the assessment of the seriousness of the predicament of sin, which is of particular interest to us. This is of interest because it impacts on the place of human participation in the consummation of all things. If human beings are fundamentally sinful, it would seem to negate any role they might play in bringing creation to God, that is, in participating in the final consummation. On the other hand, if human beings and the natural order are essentially good, or if there is some kind of balance between good and evil, then those human beings and indeed creation itself may indeed have a part to play in the transformation of all of creation.

2. Conradie, *Hope for the earth*, 25.

Conradie's key point is that 'eschatology has to retain its soteriological roots'.[3] Whilst eschatology cannot be reduced to soteriology, nevertheless the human predicaments of finitude and limitedness (which are aspects of createdness) need to be addressed within the context of the effects of sin.

I think Conradie's reminder that Christian hope is expressed primarily 'as a hope for salvation from sin, liberation from oppression and an ultimate victory over evil',[4] is a timely one for eschatologists. There is a sense in which some expressions of eschatology are in danger of becoming so caught up in the positives of process theology and of theistic evolution, that they forget about the limitedness of creation and the effects of sin.

Christian hope, suggests Conradie, is built on the negation of present sufferings, and the effects of sin, oppression and evil, which will not have the final word. Thus in Christian hope the future is 'often conceived of not so much in terms of positive anticipations and aspirations but in the negative terms of present suffering. In criticising the evil causing such suffering, people seek and hope for another and better future.'[5] Within this context the basis of Christian hope is the God of promise, and not human self-transcendence or latent natural processes. Conradie seems to be implying here a criticism of such theologians as Teilhard de Chardin, Karl Rahner, the process theologians and, even more recently, the likes of John Haught and his theology of theistic evolution. The problem then becomes for Conradie whether he can place due emphasis on human self-transcendence and the natural processes, in order to facilitate an ecological eschatology. His emphasis here is firmly on the confession of God's faithfulness, in other words, God's steadfast love (in Hebrew: *hesed*). Past and present experiences of God's faithfulness point the believer towards a future fulfilment. Conradie is careful to emphasise the importance of past and present experiences of God's faithfulness. 'Hope becomes unsteadfast unless it is based on past and present events that provide the grounds for trusting in a future fulfillment'.[6] Only the experience of how God has *already* acted faithfully allows the believer to say anything about the future. The vision of the future is cosmic in scope and is expressed as a vision of life with God, where peace and justice will prevail. Conradie points us to the prophetic tradition, which in the New Testament at least finds expression in the resurrection of Christ. The resurrection of Christ becomes the basis

---

3. Conradie, *Hope for the earth*, 259.
4. Conradie, *Hope for the earth*, 31.
5. Conradie, *Hope for the earth*, 31.
6. Conradie, *Hope for the earth*, 32.

of all hope, which longs for God's peace to be fulfilled and encompass all things. Christ's incarnation, life, ministry, death and resurrection are the fulfilment of past hopes and promises, as well as being the foundation of hope for future fulfilment. But the basis of this hope is salvation from sin, which is expressed in a number of New Testament metaphors and images, 'the great banquet, the wedding feast of the lamb, the new Jerusalem, the new earth and, especially, the coming kingdom of God'.[7]

In our recent and contemporary world the reality of the need for soteriology to play a part in eschatology is quite simply demonstrated by the current state of the world. Conradie quite helpfully describes the current ecological state of the world, and its projected future, as suffering from the ecological impact of evil. He mentions several crucial ecological problems, which have simply intensified since the main body of environmental studies began in the 1970s (the 1970s marked the beginning of a vast accumulation of evidence of regional and global environmental disasters, though clearly various studies predate the 1970s). He gives us a helpful, if somewhat depressing, summary of the environmental issues facing the global community:

- the environmental impact of global population and its threat to the carrying capacity of the earth (a function of growth of human population, consumption and the use of technology);
- the salination and toxification of soil in the first world, overgrazing and deforestation in the third world, the collapse of fishing industries due to persistent over-fishing, the virtual destruction of the rainforests and other eco-systems, often leading to desertification;
- atmospheric pollution: the threat of ozone depletion, global warming, acid rain, changing weather patterns;
- the pollution of water supplies through highly toxic forms of waste (including the problem of nuclear waste);
- the loss of biodiversity.[8]

The very survival of human beings is at stake and the fear of many is that the twenty-first century will end in an ecological nightmare. The problem is not helped by the fact that social injustice is inextricably linked with environmental degradation. Using an example from South Africa he

7. Conradie, *Hope for the earth*, 34.
8. Conradie, *Hope for the earth*, 53.

shows that the extreme poverty of many people, who are really still suffering from the effects of apartheid, under which system many people were forced to live in over-crowded conditions and on land which was incapable of supporting them all, has resulted in a vicious circle of abuse. The very land, upon which the millions of poor depend for survival, is abused as the desperate struggle for survival continues. In stark contrast the middle class and elite members of South African society still enjoy a wonderful standard of living, but which leads in turn to consumerism and wastefulness characteristic of other first world countries. Conradie recognises the links with apocalyptic consciousness here, when he suggests that beneath the surface of popular consciousness lurks a 'silent but pervasive fear for the long term future of life on this planet'.[9] This sense of fear is only increased by a loss of confidence in the capacity of technology to overcome the threat to our continued survival. Conradie's own confidence in the capacity of human beings to ensure their own survival also seems low, when he analyses some of the possible solutions to the crisis facing us. Firstly, it may already be too late, because the current and ongoing poisoning of the environment may only show its full effects in years to come. Secondly, the desire and goal of the poorest nations to achieve the standard of living enjoyed by first world countries is simply unachievable. Even if the global community were to embark on a proper redistribution of wealth, bearing in mind that the voluntary surrender of an affluent life-style seems highly unlikely, the consumerism of first world countries is simply not sustainable on a world scale. Finally, the very economic structure of first world countries depends on a high level of consumerism and pollution. But this structure has been spreading to developing nations as well, with the result that they are abandoning their simpler subsistence cultures to follow first world models of clearing land and overgrazing it. These first world models are far less ecologically sustainable than the ones they replace.

In highlighting all of this, Conradie seems to be saying that the basis of any hope rests firmly with God. This is actually in stark contrast to both Ruether and Keller, who use similar examples of environmental degradation and social injustice and violence to show that the salvation of human beings and the world lies firmly in the hands of human beings. Ultimately Conradie is more convincing, because, as we shall see, there are other limitations on human beings and the world they live in (other predicaments, to use Conradie's term) which prevent human beings, through a

9. Conradie, *Hope for the earth*, 59.

process of self-transcendence, or the cosmos itself, through its evolutionary processes, achieving salvation. The question that still remains, then, to be asked of Conradie, is whether human beings and the processes of the world have *any* role to play. As we shall see, Conradie's work certainly doesn't *emphasise* the contribution of human beings and the processes of the world, despite his clear commitment to social justice and the protection of the environment.

In asserting that Christian eschatology must be rooted in soteriology, Conradie writes: 'The proclamation of the gospel has to remain focused on the Christian vision of hope for salvation from sin, liberation from injustice and victory over evil.'[10] The danger of current creation centred theology, as he sees it, is that comparatively little attention is given to the motif of salvation from sin. What Conradie therefore calls for is the 'reintegration of the motifs of redemption and creation'.[11] He is critical of those forms of eschatology which seem to retain optimism in the processes of the world without addressing the problem posed by the reality of sin. He has in mind at this point the optimism based on evolutionary biology (especially that of Haught) and that based on information technology (specifically Dyson and Tipler). This optimism remains 'undisturbed by the radicality of human evil'.[12] What is needed is a vision in which the cosmic scope of God's acts of salvation is emphasised. In other words an eschatology is needed which addresses the reality of evil in the world. Yet, as he identifies, human sin is just one brief episode in a much larger *cosmic* history of God's creativity, nurturing and salvation. Thus, despite the importance of addressing the question of sin in eschatology, eschatology cannot be reduced to soteriology.

An interesting part of soteriology is the question of God's judgement, which Conradie deals with briefly. The context of judgement is in fact the proclamation of life in the face of evil. The Christian hopes for new life in the face of the powers of death and destruction. But this hope for new life remains inadequate and shallow, he suggests, without a corresponding judgement on the forces which threaten life: 'A prophetic theology of life will only be adequate if it announces the fullness of life and denounces threats to life. A theology of life needs to discern the idols of death.'[13] Thus Christian hope is that the forces which currently threaten life, forces of

10. Conradie, *Hope for the earth*, 256.
11. Conradie, *Hope for the earth*, 256.
12. Conradie, *Hope for the earth*, 258.
13. Conradie, *Hope for the earth*, 312.

evil, will be 'exposed and halted by a just verdict'.[14] It is only when this has occurred that true healing can take place. Otherwise the vision of wholeness offered is shallow optimism; destruction invariably continues unless it is brought to a halt by someone or something else. History can only be redeemed, he insists, if it passes through God's judgement and this judgement may indeed imply punishment. At the very least, he suggests, the powers of those who perpetrate evil will have to come to an end.

Conradie's perception of the reality of sin and its effects in the world is acute. His personal thoughts on the future of the planet make for grim reading, when he suggests that he does not think we will realistically be able to avert some major environmental catastrophes and that the world will be a much less enjoyable place to live in at the end of the twenty-first century. Yet hope, he affirms, arises from desperate situations and may even include a sense of pessimism within its framework. The stirring examples he uses of hope developed in just such seemingly hopeless situations are Jeremiah, Jesus and Augustine, all of whom realised a fundamental crisis for their society was approaching (the devastation of Jerusalem in 587 BCE, again in 70 CE and then the fall of Rome from 410 CE). Yet none of these men succumbed to pessimism, rather they 'fearlessly analysed the roots of the looming crisis, showing the devastating effects of sin and indicating the rottenness from within their own societies'.[15] Moreover, they acted in solidarity with the victims crushed by the forces of their societies and retained a vision of hope, in which the old society would be cleansed of its evil and a new community would emerge. Clearly this is also the task for eschatology today. This world must be cleansed of its evil and sin. Those who oppress, often with violence, innocent victims, must be stopped and punished ('despite the aversion that this may evoke amongst those with liberal sensitivities'[16]). This world, at least in its current state, is not yet the new creation. At least with respect to sin and evil, we can see some truth in the apocalyptic visions discussed by Körtner and Moltmann.

I think here we see Conradie's very helpful contribution to eschatology from a South African perspective. He is well placed, perhaps unlike those of us living in a predominantly comfortable first world society, to see the immediate impact of the powers of evil and destruction. Conradie's perspective is valuable to the academic world, because he allows the stark

14. Conradie, *Hope for the earth*, 312.
15. Conradie, *Hope for the earth*, 315.
16. Conradie, *Hope for the earth*, 313.

realities of elements of life in South Africa to influence his theological method. His perspective jolts people out of our sometimes idealistic vision of a new creation, allowing them to see the destruction caused by sin. Thus Conradie's eschatology reminds us, particularly those of us caught up in the very positive discussions between science and theology, that sin, evil and destruction are all-too-real forces in our world. They cannot simply be brushed over in any contemporary eschatology. There must be an acknowledgment of suffering, of pain and destruction. As I shall discuss in the next chapter, Elizabeth Johnson provides eschatology with a model, which does just that.

6.3 The predicament of finitude in time

Clearly the most confronting predicament facing human beings is the reality of their own impending death.[17] Most modern societies, Conradie reminds us, attempt to hide the reality of aging, dying and death. This is a timely reminder from Conradie, because of the overwhelming emphasis in the popular media of hiding the signs of aging, as if thereby one might slow down the process of aging and thus delay death indefinitely. Though he doesn't mention it specifically, one might think also of the sad reality that many of our aged are spirited away to old age homes and effectively abandoned by their younger relatives. For many of the aged, of course, such places can be quite life-giving, but not all enjoy the benefits of regular visits from family members.

Perhaps the greatest realisation of finitude comes when we examine the fate, not of individual creatures, but of entire species. This applies, naturally, to the human species as well. 'We are nothing more than a brief episode in the cosmic and planetary drama.'[18] The arrow of time, which itself has had a profound impact on theology and philosophy, affects even entire continents and oceans. Indeed the universe itself has a finite life span. 'The experience of finitude is therefore not limited to human beings; it has cosmic dimensions. The world itself is destined to pass away.'[19]

The predicament of finitude does not simply apply to our lives in the sense of the threat of death. There is also the notion of transitoriness, or, as Whitehead puts it, 'perpetual perishing'. This means that every human being is aware of a sense of perpetual loss. Every special moment, whether

17. See Ernst Conradie, 'On human finitude and eternal life', in *Scriptura* 88 (2005): 30–51.
18. Conradie, *Hope for the earth*, 27.
19. Conradie, *Hope for the earth*, 27.

the celebration of a birthday, Christmas or other seasons of the church's year, or simply the visit of a loved one from interstate, will inevitably pass into history and eventually be lost. Any eschatology must address the issue of transitoriness. The issue of transience finds poignant meaning in the celebration of various rituals, in which we experience the sensation of 'time standing still'. Yet these moments of eternity, in which the temporal flow is interrupted, seem only to highlight the finitude of time. The quest for eternal life in numerous religious traditions, including the Christian, is the manifestation of the deep seated human desire to overcome that sense of transitoriness.

I think Conradie here starts to address the fundamental issue for ecological eschatology. If the world itself, not just human beings, is in a continual state of transitoriness, how can we understand God redeeming the world? This is crucial, because 'the world' is not a static reality, any more than humans themselves are static beings. In fact 'the world' as some kind of static reality simply does not exist. At a cosmic level, the universe continues to expand and change. At an earthly level, the planet we live on is a vibrant, living, interconnected web of organisms, each with their own individual history and time, though these individual histories and times are connected in a complex web of relationships. Eschatology, then, must find some way of valuing *all* of this, not just human beings. Moreover, it cannot simply assert that human beings are resurrected and live (in Christ) on a transformed 'earth', because this earth alone is a sophisticated, beautiful collection of a vast array of living and inanimate things, all living in a complex, interconnected relationship with one another.

The dimensions of finitude which Conradie discusses have already been discussed in this book. However, a brief examination will help set the scene for Conradie's steps towards a resolution of some of these issues. The cosmic dimension of finitude is an issue relatively new to eschatology, just as it is to every other academic discipline. The concept that the universe has a history, and is not everlasting, is, in my opinion, one of the most interesting and challenging scientific theories in human history. 'The unpalatable truth appears to be that the inexorable disintegration of the universe is assured.'[20] It would appear at least that the cosmological consequences of entropy evoke complete hopelessness.

Conradie provides the reader with a good summary of the problem, in terms of the second law of thermodynamics, which it is helpful to sum-

20. Conradie, *Hope for the earth*, 69.

marise briefly. The first law of thermodynamics states that energy can be converted from one form to another and that, in a closed system, the quantity of energy will remain the same. However, the second law of thermodynamics holds that the conversion of energy to heat (to do some useful work) inevitably means a decrease in the *quality* of the energy. Thus the available energy will become increasingly random and disordered (the effects of entropy). In a closed system entropy is irreversible and unavoidable.

The critical question then becomes whether the universe as a whole is in fact a closed system. This is a crucial question, because the laws of thermodynamics only apply within a closed system. For example, the earth itself is not a closed system, because of the available heat from the sun. However, entropy in the whole earth-sun system is increasing. Whether, then, the laws of thermodynamics apply to the entire cosmos is a disputed question in science. It may well be the case that there will be fluctuations within the system, with regions in the universe continuing to act as open systems.

The question of finitude also applies to the earth itself. Indeed the destiny of our earth can be predicted with much more confidence than that of the wider universe. Within approximately five billion years the hydrogen in the core of the sun will be depleted. The sun's temperature will rise and it will become a red giant. Meanwhile the earth will suffer the effects of this profound increase in heat, with the oceans evaporating, all forms of life dying and then the earth itself melting. Finally the sun itself 'will eventually be reduced to a ball of burnt-out matter'.[21] It's not a very positive picture of the earth's future! This picture assumes that advanced and intelligent life forms will not have been wiped out by a nuclear catastrophe or a collision with a massive asteroid in the meantime. Yet, whilst apocalyptic world visions (at least in the framework of positive apocalypticism, using Körtner's terminology) provide a vision of a restored earth after the apocalyptic destruction, the scientific vision of the end of the world provides no such picture.

The question of finitude is, however, also much more localised than the future destruction of the earth. Setting aside the fascinating question of the survival of the human race (which Conradie suggests will disappear from the earth sooner or later, if the history of our prehistoric ancestors provides any kind of guide), the question of the *current* loss of individu-

21. Conradie, *Hope for the earth*, 73.

al species provides eschatology with yet another example of finitude. Of course not only have species become extinct over the entire course of the earth's evolutionary history, but this process has been enormously accelerated. The current loss of biodiversity means that thousands of species are being driven to extinction every year.[22]

Finitude remains the ultimate hurdle in Conradie's thinking, though interestingly he suggests that the predicament of sin and evil should take priority over the question of finitude in being the chief focus of Christian eschatology. His reasoning is that speculation on life after death and the question of finitude potentially becomes too distant from everyday concerns. On the other hand, a theology focussed on questions of sin and evil, particularly the present ecological threats to life here and now, seems far more appropriate and less esoteric. Conradie enters here the debate about whether hope for the human race, the earth and the cosmos in fact detracts from or adds to a commitment to ecological concerns here and now. He accepts that many people are prepared simply to accept finitude, both their own and also the ultimate finitude of the earth. However, the reality of finitude neither dampens their spirit of hope for the future nor prevents them from rejoicing in the gift of life in the present. His South African experience suggests that the spirit of hope expressed by those involved in the long term struggle against the oppression and violence of apartheid was not really influenced by 'speculations about the long term future of the cosmos'.[23]

Whilst the point is accepted, I was encouraged to note that, in the end, Conradie returned to a position which expressed how helpful a long term vision is in providing inspiration for one's present existence. 'When hope disappears, life itself soon becomes meaningless.'[24] In other words, an encompassing vision is in fact needed, because *anthropocentric* motivations for an environmental praxis cannot give us an ultimate vision of hope. The foundation for an environmental ethos depends on a broader vision, otherwise nihilism can prevail and destroy all motivation for caring for the

22. The fate of the dinosaurs continues to be an interesting sub-theme of my thinking, in that respect. As I've already noted in my critique of Jürgen Moltmann's theology, their extinction allowed other species to develop, including mammals, which eventually led of course to the rise of human beings. Yet their extinction remains a dilemma for eschatology. Do we write them off as having had their time? Or does eschatology provide us with other ways of coming to terms with the extinction of the dinosaurs and other species?
23. Conradie, *Hope for the earth*, 77.
24. Conradie, *Hope for the earth*, 79.

earth. The temptation to despair is closely connected with the question of finitude. Using Conradie's thoughts here, I would suggest that ecological ethics can greatly benefit from an eschatology, which is environmentally friendly. As Conradie points out, in agreement with Ruether at this point, there must be a connection between this earth and the new creation; there needs to be some sense of continuity between my earthly body and my resurrected body, between the cosmos and the eschaton, otherwise we become like those who preach 'damaged' apocalypticism. We may even hope for the end of the world and retard efforts to clean up the environment, because these efforts might delay the end.

Conradie also makes a connection, with Moltmann, between the predicament of sin and evil and the predicament of finitude. If death is the end of everything, then *ultimately* justice is impossible for those who have died. Even their memory will eventually fade, at least from human memory (though *God's* memory is a different matter, as I shall discuss). Thus the predicament of finitude becomes a stumbling block for victory over evil. But the same is true of the reality of transitoriness. 'If one has lost certain opportunities for good, if one's best years have been ruined through pain or injustice, this can never be given back. There will be a lasting sense of loss for the time that has gone by.'[25] Whilst some might see past injustices as a powerful source of inspiration for pushing towards liberation in the future, Conradie is wary of such approaches. The reign of God, he reminds us, remains *God's* reign, and cannot become or be confused with a human driven historical movement. Certainly, I would suggest, too often the memory of past injustices and victims has been manipulated to justify horrific acts during a process of revolution. And, of course, even when past injustices function as a source of inspiration for future action, the innocent victims of the past are still left without justice. No achievement of future goals can bring justice to the victims of the past. Nevertheless, as I will show later in the book, memory should not be underestimated as a powerful source of inspiration for resistance against oppression.

In Conradie's theology justice for the past is linked with the hope for vindication in life after death. I would articulate what Conradie simply implies here, that for this reason any theology that precludes some kind of eschatological vision is inadequate because of the question of justice. 'Without judgement there is no justice. Without resurrection, justice cannot be complete. The Christian hope for a final judgement is born from

25. Conradie, *Hope for the earth*, 319.

this sense of justice.'[26] Ultimately then, as Conradie rightly keeps reminding us, salvation comes from God.

Suffering within creation is not always the result of human sin and evil, it is in fact an inherent part of nature, Conradie suggests. Here, as he freely acknowledges, Conradie moves beyond the dominant strand of teaching in the Christian tradition, in which there is a causal connection seen between death and human sin. But 'suffering, violence, decay, death, and the extinction of species was an integral part of nature from the very beginning'.[27] Whilst Conradie continues to insist that Christian theology should maintain its focus on sin and its effects, all suffering is not the result of sin—some suffering can be explained as being the result of temporal finitude and the distinction between the two needs to be acknowledged. Although an integrated eschatological response to the predicaments of finitude and evil needs to be developed, the two predicaments ought not to be conflated.

6.4 The open or closed universe

To return to an earlier problem, if the question of the open or closed nature of the universe itself is difficult enough to answer scientifically, theologically it is surely impossible. A question which Conradie therefore examines is whether there is a cosmic dimension of hope, which might provide an alternative vision of the future to the one proposed by most scientific theories about the end of the cosmos. In the face of ecological despair, he suggests, a wealth of purely metaphorical images of hope will not be enough to sustain an environmental praxis. His warning is well heeded, that 'Christian eschatology dare not offer cheap forms of hope'.[28] This leads Conradie to an examination of some of the responses to the crisis on entropy or a cataclysmic end to the universe, from those contributing to the contemporary dialogue between science and religion. The predicament of cosmic finitude seems to dominate the responses. Is there a theological solution to the question of entropy and the final consummation of the universe, which is nevertheless not simply metaphorical or, as Moltmann would put it, involves history being subsumed by eternity?

A detailed examination of Conradie's work on the question of cosmic dimensions of hope is warranted here, because the increasingly significant

26. Conradie, *Hope for the earth*, 324.
27. Conradie, *Hope for the earth*, 326.
28. Conradie, *Hope for the earth*, 221.

science and religion dialogue has featured a number of contributions in this area. An important theologian for Conradie, whose work, though, he approaches with a certain amount of caution, is the American writer John Haught. The key concept, which Haught develops at some length, is that nature itself (by which Haught actually means the cosmos) shows inherent promise. There is a fundamental openness within creation and it is intrinsically less predictable and quantifiable than earlier scientists assumed. Moreover, the universe has a dynamic and ever-changing history, which means that in some ways the scientists who attempt to describe it are as much story tellers as objective observers. Using Haught's approach, Conradie reminds us that the history of the cosmos has also been characterised by increasing complexity, 'advancing towards more sentient, conscious, and eventually self-conscious forms, culminating temporarily in the emergence of the human species'.[29] This increase in complexity is one of the signs of inherent promise in the universe; its conditions were delicately configured as to 'make the eventual emergence of life and consciousness a possible and even probable development'.[30] There is then, according to Haught, a sense of a cosmic adventure, in which restlessness, suspense and mystery abound, leaving the observer with the question of how it is all going to end. For Haught, the direction being taken is towards increasing consciousness and beauty, which is, however, only recognised if one regards the cosmos as being full of promise. Guided by eschatology, and the stories of the Christian faith, the Christian discerns a pattern of promise in the evolving universe, which will blossom into a new creation.

However, Haught recognises the limitations of nature, which is 'good' rather than 'perfect'. In its beauty, however, we see a future fulfilment and beauty, which is yet to manifest itself. But the present struggle within creation and the present experience of beauty, suggests that creation may be involved in the birth of something truly new. As Conradie puts it, 'The whole cosmic process is a purposeful struggle to realise an unprecedented promise of new being.'[31]

Two current scientific theories also contribute to Haught's sense of cosmic promise: the theories of chaos and complexity. Without examining the finer detail of the theories, the pioneering work of Ilya Prigogine on 'non-equilibrium thermodynamics' suggested that 'when a system is

29. Conradie, *Hope for the earth*, 226.
30. Conradie, *Hope for the earth*, 227. See John Haught, *Mystery and Promise: A Theology of Revelation* (Collegeville, Minnesota: The Liturgical Press, 1993), 154.
31. Conradie, *Hope for the earth*, 231.

far from equilibrium, a new complex form or order sometimes emerges and achieves a stable form'.[32] These systems are created 'through the entropy-producing dissipation of energy in open systems'.[33] Thus, order and complexity emerge out of chaos. Chaos theory, on the other hand, describes the apparent random behaviour of systems, which are affected by tiny uncertainties in the initial conditions. These chaotic systems can also be affected by what are called strange or chaotic attractors, which in turn increase the unpredictability of the system. Still other studies have identified the phenomenon of self-organisation within complex systems, where complexity may emerge unexpectedly from a state of thermodynamic turbulence. This complexity cannot be explained by the influence of the constituent parts of the system.[34]

Haught uses these two theories to emphasise the unpredictability of chaotic systems, and their apparent indeterminacy. Moreover, he emphasises the possibility of 'top-down' causation, because of the fact that complexity may emerge in complex systems, beyond what one would expect from their constituent parts. Conradie finds it significant that, 'once a material system reaches a certain level of complexity . . . then genuinely novel properties and processes may emerge, which cannot be explained by the material properties of the components themselves'.[35] These theories of chaos and complexity are therefore anti-reductionist and holistic. Haught suggests therefore that entropy does not necessarily signal the heat death of the universe, but the opening up of new possibilities, because 'the interplay between chaos and ordered complexity illustrates both the reliability and the openness to surprise inherent in nature'.[36] This is also consonant with the Christian understanding of God. Thus the future is not pre-determined by the second law of thermodynamics. Rather nature contains within itself the possibility of a genuinely novel future; this possibility resonates with the Christian understanding of God and his purposes for us.

But Conradie is not convinced by Haught's work. Here we see Conradie's central concern that hope for the future not be linked with the long-term future of the earth and the dimensions of time and space. His central concern, as I shall discuss, is that any eschatology must be theocentric.

32. Conradie, *Hope for the earth*, 234. See Ian Barbour, *Religion and Science: Historical and Contemporary Issues* (San Francisco: Harper & Row, 1997), 180f.
33. Conradie, *Hope for the earth*, 235.
34. Conradie, *Hope for the earth*, 236.
35. Conradie, *Hope for the earth*, 239.
36. Conradie, *Hope for the earth*, 241.

I suspect *this*, rather than necessarily his stated concerns with Haught's theology, remains the root of his ultimate rejection of any kind of evolutionary based hope. Nevertheless Conradie does raise some fundamental questions about Haught's use of chaos and complexity to discern a sense of promise in nature. Firstly, chaos theory does not necessarily support a theory that the entire environment is not causally determined. Unpredictability within a small part of the whole system does not equate to indeterminism of the whole, even if chaos theory put *limits* on how much determinacy can be established. Secondly, the theory of self-organisation and increasing complexity does not avoid the issue and inevitability of human finitude and death. Likewise, I would add, the finitude of the cosmos is not avoided by these emerging theories of increasing complexity. If the universe as a whole is a closed system, then pockets of increasing complexity will but delay the inevitable running down of all energy.

Conradie, then, finally rejects Haught's notion of the promise of nature, as well as, not surprisingly, Frank Tipler's[37] notion of an Omega Point, where all knowledge will be drawn to a single mathematical point and be able to comprehend the entire history of the universe at once. Interestingly Conradie's rejection of Haught's work rests on his insistence that salvation cannot come through any process of nature, however promise-filled nature might be. In fact, as Conradie notes, even Haught does not see the promise of nature as being identical to its fulfilment. As Conradie puts it, 'A blossoming youth provides no guarantee against old age. All events, all forms of life, even the cosmos as a whole remain ultimately futile.'[38] Indeed Haught and Conradie agree that it is God's action which provides an ultimate hope for the cosmos. For Haught this hope finds expression in the concept of the world being received into God's 'everlasting experience' and attaining 'objective immortality'.[39] Christian hope, Conradie insists, remains rooted in God's faithfulness, not in the unfolding of the physical processes. Indeed, he argues, the finitude of the cosmos should not pose a threat to eschatology. God's faithfulness ensures that the entire history of the universe, beyond death and destruction, will be taken up into God.

37. See Frank Tipler, *The Physics of Immortality: Modern Cosmology, God, and the Resurrection of the Dead* (New York: Doubleday, 1994).
38. Conradie, *Hope for the earth*, 252.
39. John Haught, *The Promise of Nature: Ecology and Cosmic Purpose* (New York/Mahwah, NJ: Paulist Press, 1993), 133. See also John Haught, 'Ecology and Eschatology', in *And God says that it was good. Catholic theology and the environment*, edited by D Christiansen and W Granzen (Washington: United States Catholic Conference, 1996), 47–64.

Interestingly Conradie does not explore here in depth those other theologies (apart from Haught) which emphasise the continuity between the old and new creation, but which rely on God as the ground for this continuity. Part of the issue seems to be the fact that Haught wrote some of his best theology on this issue in the same year that Conradie's chief work on eschatology was published.[40] It is Haught's more recent work on God as the Absolute Future, which gives his theology the transcendence often lacking in processive approaches to divine action in general and eschatology in particular. He suggests that God is not a timeless reality, located somewhere 'up above', but is in the future 'up ahead'. He argues against the reductionist, pessimistic scientific materialist assertion that nothing new ever occurs in creation and that there is no new creation. A metaphysics of the future is an alternative to a metaphysics of the past or of the eternal present. When God withdraws his presence, he withdraws into the future. This concept allows an ultimate explanation of what Haught describes as the raw ingredients of Darwinian evolution, that is, contingency, lawfulness and temporality. Contingency is nature's way of being open to the future, to new creation. It is the way 'in which the cosmos breaks out of complete subordination to habitual routine and opens itself to the future'.[41] In the case of law, Haught suggests that lawfulness and predicability give structure to the process of evolution. They are needed to keep the natural processes 'continuous enough to avoid decaying into utter caprice'.[42] With regard to time, the future allows the pushing of the present into the past and creates a temporal sequence of moments in which evolution can occur.

If God as described by Haught is both kenotic love and the power of the future, then God's purpose for the world and interaction with the world is a much broader concept than any understanding of design. Haught suggests that God's purpose includes a promise for the whole of reality, the physical cosmos and what might be recognised as the 'entire sweep of its evolution'.[43] In letting creation be, that is, in conceding to the world 'its own autonomous principles of operation'[44] God does not, however, lose intimacy with the world. Nor does he abandon the world in some way. The

40. Conradie's work, *Hope for the earth*, was published in 2000, as was John Haught's work, *God After Darwin: A Theology of Evolution* (Boulder, Colorado: Westview Press, 2000).
41. Haught, *God After Darwin*, 102.
42. Haught, *God After Darwin*, 103.
43. Haught, *God After Darwin*, 108.
44. Haught, *God After Darwin*, 114.

Creator's capacity to influence the world is 'made manifest paradoxically in the vulnerable defencelessness of a crucified man'.[45] And his promise is that each moment of the history of evolution will be 'taken perpetually into, and preserved everlastingly in, the boundlessly redemptive future that faith names as God'.[46] This includes any 'heat death' which may be awaiting the physical universe. Yet the future of the cosmos remains open. The movement within the cosmos towards increasing complexity and the fact that the physical universe always held at least 'the promise of emerging into life'[47] suggests that it is not necessary to take a pessimistic view of the future of the cosmos.

Rahner's notion of God as the absolute future would assist here in providing a model of how this world might indeed be continuous with God's new creation, without relying solely on notions of cosmic evolution or nature's promise. For Rahner the consummation of all things is the *gift* of God. Yet in Rahner's understanding matter is never to be regarded as transitory in the spirit's path to consummation, but rather endures 'as an intrinsic element of the spirit and of its history even in its consummation'.[48]

A fascinating short story by Isaac Asimov addresses this very question and provides a solution to the question of entropy, albeit one which does not naturally sit alongside the Christian position.[49] In his story *The Last Question*[50] Asimov puts the question, 'How can the net amount of energy of the universe be massively decreased?' into the thought processes of one of his characters, who in turn asks it of a sophisticated computer. The computer simply prints the reply, 'INSUFFICIENT DATA FOR MEANINGUL ANSWER'. Millions, billions and then trillions of years pass, with each generation of Artificial Consciousness (AC) gaining in sophistication and intelligence, until the final embodiment AC is able to answer the question of entropy by saying, 'LET THERE BE LIGHT!'

45. Haught, *God After Darwin*, 113.
46. Haught, *God After Darwin*, 115.
47. Haught, *God After Darwin*, 117.
48. P Phan, *Eternity in Time: A Study of Karl Rahner's Eschatology* (London & Toronto: Associated University Presses, 1988), 196.
49. The story in fact avoids the difficult theological questions by providing a solution based on the idea of recurring universes.
50. Isaac Asimov, 'The Last Question', in *Nine Tomorrows: Tales of the Near Future* (London: Dennis Dobson, 1963), 177–190. This short story first appeared in *Science Fiction Quarterly*, November (1956).

6.5 The predicament of finitude in space

The third predicament, which Conradie identifies, is the predicament of human limitedness in space, which has important consequences for the limit of human power and knowledge. We are limited, however much we as a species might attempt to transcend those limits, by space. There is only so much space we can exert an influence on, however much we might extend our influence to those limits. There are also limits, set by our very humanity, to what we can know. Unlike God we cannot enjoy any measure of omniscience—the very physical nature and limitations imposed by the size and capacity of the human brain determine this. Yet we cannot help asking questions about those things which transcend us. The predicament of limited power and knowledge leads to an anxiety about the future and the powers beyond one's control. In fact it leads, often, to a search for God and an interest in God's transcendence.

In this context Christian hope is that the limits we experience may be transcended in the presence of a transcendent God. Eschatology then has the task of exploring the relationship and dynamics of God's immanence and transcendence, particularly as it relates to the finitude of human knowledge. Can we talk about human beings participating in God's eternity and enjoying an encompassing vision, which transcends the limited knowledge we now have? In the development of Christianity, the distinction between God and human beings was (and still is) understood to be an ontological separation. The particular expression of Christian hope in this context is then the quest to return to God's presence, and thus experience the beatific vision.

However, the issues arising from this eschatological vision are its limited nature (its emphasis on the individual's relationship with God, but not the importance of the earth in that equation) and its emphasis on reaching eternity. God's transcendence is emphasised over and against his immanence, placing an undue emphasis on reaching life after death. Eternity is then seen as diametrically opposed to time, or, to use Moltmann's turn of phrase, eschatology is transposed into eternity. Dialectical theology, particularly as developed by Karl Barth in the first part of the twentieth century, tried to make a fundamental distinction between Christian eschatology and faith in the historical process. The distinction between God and the world was highlighted. Barth's eschatology had an overwhelming emphasis on God's transcendence, that is, on the vertical dimension of eschatology. So much so, that it is criticised for being strangely a-historical and, ironically, lacking a focus on the future: 'The horizon of the future is swallowed up by the eternal blitzing from above. Eschatology is trapped in an eternalised present experience.'[51]

In response to Barth several theologians, including Moltmann and Pannenberg, developed an emphasis on the transcendence of the future. As I explored in the work of Moltmann, the ontological priority given to the future allowed an understanding of God coming from the future. This notion also allowed for history, including the future, to be given room within the eternity of God.

Having explored Moltmann's work, and that of Pannenberg, Conradie then moves to an examination and criticism of those theologians who over-emphasise God's immanence. These eschatologies, he suggests, which include Haught, Arthur Peacocke and Teilhard de Chardin, look for a promising future within the material history of the cosmos itself. They are followed in this by some eco-feminist theologians (as I discussed in section one) who emphasise God's immanence. Given the general thrust of his work, it is not surprising that Conradie finds these theologians generally unsatisfactory, because any source of hope cannot be located primarily in God's immanence, but in God's transcendence. Any eschatology, which is based on an emphasis on immanence cannot remain hopeful in the face of the finitude of the cosmos itself. An emphasis on God's transcendence remains vital for Christian eschatology, because 'if there is any hope beyond the deaths of human persons, the human species, life on earth, the planet itself and the universe, then this hope can only lie in being taken up into God's eternal presence'.[52]

Here we come now to the central plank of Conradie's argument, reiterated throughout his work, that the 'destiny of the cosmos remains in God's hands'.[53] The future lies with God. Obviously this assertion rests at a very basic level on the belief that God is more than a human construct. It also rests primarily on God's transcendence, over and above his immanence. The final consummation of the cosmos will be an act of God. Contrary to what Körtner was exploring in his work on the absurd, and the thought that the end of the universe would also be the end of any human memory of God, Conradie expresses the hope that the final victory belongs to God, *even if* the universe itself should end. What eschatology portrays therefore is simply the future with God: 'The cosmic dimensions of space and time are transcended by God's presence in eternity. God's eternal presence per-

51. Conradie, *Hope for the earth*, 197–198.
52. Conradie, *Hope for the earth*, 219.
53. Conradie, *Hope for the earth*, 336.

meates, embraces and governs the present world. The key to the future of the cosmos remains in God's hands.'[54] The Christian hope is that everything, the entire, vast history of the cosmos itself, will be taken up into the loving presence of God.

6.6 Theocentric focus as a response to the three predicaments

Theocentrism, then, provides the key to the theological continuity between the cosmos and the eschaton. This continuity is primarily a continuity in God's self, brought about through God's steadfast love of his entire creation. Therefore Conradie responds to all of the predicaments of the human (and cosmic) condition by insisting that any hope for 'the resurrection of the body and the new earth is only possible on the basis of a theocentric focus'.[55] The task for any eschatological vision is then to articulate the nature of God's relationship with the world and of God's immanence and transcendence with respect to the cosmos. This is a difficult task for Conradie, because, like Moltmann (one of his favourite eschatologists) his focus is indeed theocentric, which means discussing the role of human beings and creation itself in the eschatological vision becomes so much more difficult. This is possibly the only weakness in Conradie's otherwise inspiring vision, which I shall now examine.

Before developing the idea of multiple dimensions (or, more accurately, two worlds) as the basis of his theocentric vision of the eschaton, Conradie firstly discusses three concepts which explore the question of continuity between the cosmos and the eschaton: God's cosmic memory; a cosmic pilgrimage; a homecoming banquet. The notion of 'God's cosmic memory' has been explored by several theologians. Whitehead[56] developed the notion that whatever happened in history could not be undone, but that every event makes a difference in the future and that God is (by nature of his dipolar relationship with time and history) influenced by every event in history. Every moment is inscribed in God's memory and taken up into God's existence. Moreover, this memory is active: 'When something is remembered it comes alive again in the present'.[57] Nevertheless Conradie suggests that the great limitation of this understanding of

54. Conradie, *Hope for the earth*, 338.
55. Conradie, *Hope for the earth*, 339.
56. Conradie, *Hope for the earth*, 342. See Alfred Whitehead, *Process and Reality*, edited by D Griffin and D Sherburne (New York: The Free Press, 1978), 523.
57. Conradie, *Hope for the earth*, 344.

memory is that creation no longer remains independent of God. The distinction between the Creator and the creation becomes blurred, such that creation is absorbed into God's eternity.[58]

The idea of a 'cosmic pilgrimage' is that of a sacred journey, the aim of which is to be enriched by many experiences along the way and to act for the sake of others.[59] As the journey continues it is filled with a rich recollection of a lifetime of experience. The history of the cosmos may be understood in this context, as a kind of pilgrimage. The focus of the journey is to live each moment to the fullest, but with a careful balancing of past, present and future. Conradie suggests that Robert Pirsig's book, *Zen and the Art of Motorcycle Maintenance*[60] may be understood in this context, that there is a need for integration of past, present and future with every footstep. Yet, Conradie insists, this notion of the cosmic pilgrimage shares the weaknesses of the idea of God's cosmic memory, namely, what happens to the history of the cosmos in the end?

Finally the concept of a 'homecoming banquet' is based on the parable of the prodigal son, who returns home to the embrace of his father. It addresses the problems of sin, evil, injustice and alienation. The father calls his son home and waits for him to return. While he is still on the way he runs to embrace him and ensures that the rest of the villagers do likewise. Then there is a festive banquet for the entire village. The image is earthy and concrete, Conradie suggests, and the banquet takes place in a spirit of gratitude, hospitality and joy. The vision of the eschaton as a homecoming banquet emphasises the continued place of creation in the eschaton. With the image of the elder son it also provides for the possibility of further reconciliation, with the unfinished business yet to be resolved. Conradie also likes this image because it allows us to deal with the reality of sin and alienation. However, it remains an incomplete image by itself, because it still doesn't provide *how* we might imagine this cosmic homecoming banquet in the light of our particular history and indeed that of the whole cosmos.

58. Conradie, *Hope for the earth*, 344.
59. See Ernst Conradie, 'Stewards or Sojourners in the Household of God', in *Scriptura* 73 (2000): 153–174.
60. Robert Pirsig, *Zen and the art of motorcycle maintenance* (London: Corgi Books, 1974).

6.7 Multiple dimensions (or two-worlds)

Ernst Conradie writes that the model of multiple dimensions allows us to understand the relationship between space-time and eternity and so comprehend the presence of God's transcendence within the history of the cosmos. God's transcendent power, which goes far beyond our limited power, is the foundation for cosmic hope. Yet the predicament of the finitude of human knowledge and power prevents us from recognising the transcendent God. Conradie suggests that the problem in obtaining knowledge about God does not, in fact, lie in the ontological distinction between the Creator and his creation, or even in the limitations of human knowledge. Rather it lies in the separation caused by sin, not the limitations of time and space which are an integral part of our existence as created beings. Consequently salvation can never be described as a human achievement—it comes from God and is God's gift to us. It is for this reason that Conradie concentrates so significantly on soteriology in his work. However, eschatology cannot be reduced to soteriology, but rather has to return to the doctrine of God and provide a model of God's transcendence and immanence, that allows for a theological vision of hope for the earth.

The balancing of God's immanence and transcendence is not an easy task when related to eschatology. If it is true that Keller and Ruether tend to come down on the side of God's immanence (with unfortunate consequences for their respective eschatologies), then Conradie definitely comes down on the side of God's transcendence. Whilst he acknowledges the potentially devastating consequences of over emphasising transcendence, thereby exaggerating the ontological separation between God and his creation, his work nevertheless favours a transcendent model of God's interaction with the world. Conradie argues that the pendulum has swung too far in favour of an immanent model of God's interaction with the world, and suggests that almost all ecological theologians have followed Moltmann in proposing a panentheistic notion of God's interaction with the world. He admits that the emphasis has the good effect of ensuring that 'the world of nature itself is taken up in God's loving care and creative love'.[61] Nevertheless his own model of God's providence relies completely on God's transcendence, because, as he keeps insisting, 'hope for the earth can ultimately only be based in God'.[62] An ecological eschatology must, therefore, be based on God's transcendence, but with a detailed and de-

61. Conradie, *Hope for the earth*, 355.
62. Conradie, *Hope for the earth*, 355.

veloped concept of how we can discern or understand the presence of the Transcendent within the Immanent.

Conradie is right to pursue this line of thinking. Having been alerted to the dangers of overemphasising God's transcendence, and the perception of a separation between the earthly and the divine which can result, theology is able to explore how God's transcendence might encompass God's immanence. Indeed for God to be truly immanent, in the sense of being present in every time and place, it is necessary for God to be transcendent. Transcendence and immanence are not so much opposite each other as complementary.

With this crucial point in mind, and having demonstrated the limitations of some other previous theological models, Conradie then comes to the climax of his work: the discussion of a 'dimensional' understanding of God's transcendence. His ideas are premised on the concept that time is an integral part of the created order, that God is both the Creator of time and space, and therefore 'must exist beyond the finite dimensions of both space and time'.[63] In other words, eternity exists beyond the dimension of time and encompasses time. It is therefore neither 'timeless identity' nor 'endless time' but rather, I would suggest, belongs to a separate category altogether. To understand how the eternal God might encompass time, Conradie develops the notion of a 'depth' dimension, which I shall now explore in more detail. Conradie develops here Luco van den Brom's notion of divine omnipresence and his exploration of the concept of a 'higher-dimensional interpretation of the concept of divine omnipresence'.[64]

In Conradie's two-world model eternity is conceived as another dimension, a 'depth' dimension, 'beyond the edges of the space-time continuum'.[65] Thus it is beyond the continuum which we human beings inhabit. It is, I would add, beyond the realm of our normal experience, although, as I shall explore, there may indeed be moments of what Polkinghorne calls intersection; the celebration of the Eucharist is a good example of this. Importantly this dimension (or realm) of eternity includes time and space. There is no implication of the erasure of temporality. This deeper dimension of eternity is three dimensional: 'Eternity implies neither timelessness nor spacelessness'.[66] Thus it includes the space-time con-

63. Conradie, *Hope for the earth*, 358.
64. Luco van den Brom, *Divine Presence in the World: A Critical Analysis of the Notion of Divine Omnipresence* (Kampen, The Netherlands: Kok Pharos, 1993), 305.
65. Conradie, 'Resurrection, Finitude and Ecology', 289.
66. Conradie, *Hope for the earth*, 360.

tinuum, but also allows for the transcendence of space and time so that, for example, past and present events may be experienced together.

To aid the reader's comprehension of this notion, Conradie uses the rather helpful story about 'Flatland', based on the work of Edwin Abbott, *Flatland: A Romance of many Dimensions*,[67] in which the inhabitants of a two dimensional world are challenged to think in terms of three dimensions. They are completely unable to do so, until they hear a story about the limited thinking of the inhabitants of a country known as 'Lineland', where the inhabitants are so simple they only exist in one dimension. This story gets them thinking and they wonder whether perhaps the notion of a three dimensional land is not as ridiculous as they first thought. By extension in our thinking the concept of another dimension, or another depth, may simply be hard for us to grasp because we are confined in our thinking to the concept of a three (or four, including time) dimensional world.

What Conradie suggests is that our notion of eternity can be understood as a 'dimensional notion of God's transcendence'.[68] In this understanding God may be both transcendent and yet also omnipresent. Conradie admits that this understanding of multiple dimensions may have rather profound implications for our wrestling with the classic theological issues, such as the coming of Christ and the intermediate state. Yet, if we accept Conradie's model for the moment at least, there are immense benefits in his suggestions. As he points out, together with Polkinghorne[69] and van den Brom[70], it certainly helps our understanding of the resurrection appearances. Conradie writes that the resurrected Jesus appeared in a bodily form that the disciples could recognise. Jesus appears to the disciples and does very physical things, such as showing them his physical wounds and eating fish. Yet the risen Jesus is not reducible to the body of the crucified Jesus, or a 'resuscitated corpse'. The eschaton breaks into this world—it is not only the case that this world is being transformed toward the eschaton.[71] This provides us with a fresh perspective on the concept of the 'already but not yet' paradox of the eschaton, that is, that the eschaton is already here, but has not yet arrived. If the eschaton has indeed arrived,

67. Edwin Abbott, *Flatland: A romance of many dimensions* (London: Seeley & Co, 1884).
68. Conradie, *Hope for the earth*, 364.
69. Polkinghorne, *Science and Christian Belief*, 170.
70. van den Brom, *Divine presence in the World*, 305–307.
71. Conradie, 'Resurrection, Finitude and Ecology', 289.

in another dimension, it will break into our world, even though our world is not yet fully transformed.

This dimensional notion of God's transcendence, Conradie maintains, provides us with a thoroughly theocentric vision of life, which transcends any anthropological or cosmological visions. Because the finite is incorporated into the eternal, the eschaton is then understood as the full participation of the finite in the eternal. Heaven, in this model, is a dimension which transcends the earth, or rather the entire cosmos, but also includes it. The eschaton is then the transformation of the cosmos into the heavenly realm. Importantly, Conradie wants to avoid any return to the escapist notions of human beings departing this earth and going 'up' to heaven, leaving behind their body and indeed the whole of creation. The heavenly realm encompasses *all* of creation.

The issue which still faces Conradie, as it does any theologian attempting to write a thoroughly contemporary eschatology, is *how* this transition from time to eternity takes place. He reminds us that the question of finitude still looms large. Earthly time is essentially transitory. Every moment is a moment of loss and change. Conradie suggests that we need a model which incorporates both simultaneity and duration. Simultaneity is the understanding of an 'eternal now', which has the attributes of eternity; in a moment we experience the intersection of past, future and present. But this sense of simultaneity always remains fragmentary, and does not allow for a sense of unity or duration. The problem of duration, on the other hand, at least as it has been traditionally understood, is that is has been popularised as 'endless time'. There is a limitation to this understanding of duration, because it becomes impossible 'to heal the past and to structure the future'.[72] This notion of eternity as timelessness limits any possibility of change in the future, because, as the term suggests, the future no longer exists as an independent category. Yet, as Conradie correctly and helpfully points out, 'a cosmos that is transformed in the eschaton in a way which excludes material change would be completely foreign to us and would provide little hope or comfort'.[73]

72. Conradie, *Hope for the earth*, 372.
73. Conradie, *Hope for the earth*, 373.

6.8 Material inscription

To address the issue of how we might include both the aspects of simultaneity and duration, and to address how time might make the transition to eternity, Conradie explores the concept of 'material inscription'. 'The history of the cosmos, this "cosmic pilgrimage", is *inscribed* in the eschaton.'[74] Nothing that has been inscribed can be removed or pass away—it endures. These inscriptions then provide the building blocks, as it were, for life in the eschaton. Every moment, therefore, is of eternal significance. This includes, Conradie is quick to point out, moments of ecological action and preservation. Indeed caring for the earth and its creatures becomes absolutely crucial, because these are the building blocks for the new creation.

Conradie admits that the models of multiple dimensions and 'material inscription' have no soteriological thrust and are not liberating in and of themselves. In order to allow for the healing of history as a whole, there has to be the possibility of the retrieval of the past. This hope includes the hope that God would sanctify all living things, all moments, all history, when they are inscribed. There has to be an eschatological healing of history: 'The soteriological thrust of Christian hope is that the powers of death and destruction will be exposed, judged, and conquered through God's love. The primary scandal is that this has taken place through the cross of Jesus Christ.'[75] This notion of the healing of history must be included, otherwise moments of pain and suffering would endure eternally as well, and the vision of eternal inscription would in fact be one of endless suffering and torment. But God's eternal and loving presence will permeate each part of the cosmos, bringing wholeness and sanctification. In fact Conradie allows for a rather lovely notion of us being given sufficient space and time, within eternity, to address the rather limited building blocks of our past 'in order to heal the brokenness of our past and to celebrate the joy of forgiveness, with Christ, and together with the community of all creatures in the presence of God'.[76] The tragic and terrible moments of history will not simply be forgotten, but rather healed through Christ's redemptive work. God's reign, Conradie reminds us, also includes judgement. The many evils which have characterised human existence on earth will be confronted. He even allows for the notion that those who might

74. Conradie, *Hope for the earth*, 373.
75. Conradie, 'Resurrection, Finitude, and Ecology', 293.
76. Conradie, *Hope for the earth*, 375.

have died violently or prematurely will have the opportunity to complete their life stories.

This notion of 'material inscription' might appear at first to be similar to the concept of the history of the cosmos entering into God's eternal cosmic memory, as developed in process thought, with the consequent lack of embodiment which that idea suggests. However, Conradie is also keen to develop the notion that the coming alive involved in this process of material inscription 'is more than a mere memory of a distant past. It is an embodied celebration in which everyone, inscribed in the history of the cosmos, can participate in God's presence.'[77] Thus the term 'inscription' ought not to be misunderstood as simply referring to a notion of writing, nor as referring to the inscription only taking place in the mind of God (as in process theology). The cosmos is *materially* inscribed, with all of the bodily activities which this implies. Conradie challenges us to think about the notions of physical activities within the eschaton, including the very material delights of 'eating and drinking, of swimming and dancing, of sexual pleasure, of praise and worship.'[78]

One of the wonderful contributions which the notion of the 'two-realms' makes to eschatology is that it overcomes the issue of scientific predictions about the end of the world. Whatever scientific scenario may prove to be correct, whether a heat death or a cataclysmic final 'crunch', finitude can be accepted as an inherent and integral part of creation. Christian hope, as Conradie points out, is not based on the endless continuation of the universe *or* on the fulfilment and consummation of the universe within the realms of time and space. Rather our belief is that the entire cosmos will be transformed in God's transcendent presence: 'This hope in God cannot be falsified by new scientific theories, precisely because it pertains to "higher" dimensions that are not accessible to the physical sciences.'[79]

The criticism that can be made of Conradie's notion of material inscription is that it appears to be a little too automatic. It's almost as if every moment is inscribed, as it occurs, whether human beings and the cosmos want them to be inscribed or not. The notion of human beings and the cosmos playing some kind of *active* role in this process of inscription is not included. Conradie does allow for a sense of human beings *preparing* themselves for their eternal destination. He also suggests that this prepa-

77. Conradie, 'Resurrection, Finitude, and Ecology', 295.
78. Conradie, 'Resurrection, Finitude, and Ecology', 294.
79. Conradie, 'Resurrection, Finitude, and Ecology', 290.

ration requires reparation, and in a context of economic injustice, also restitution and a transformation of the structures of society'.[80] But the activity of human beings and the very processes of the cosmos need to be valued for their participation in the redemptive process. In the next section I shall examine the eschatology of Elizabeth Johnson and Alexander Schmemann and suggest that their work can be used as a model, to understand how human beings in particular, and the whole created order more generally, might *participate actively* in this process of 'material inscription'.

Whether Conradie's models are sufficiently *teleological* remains the question. Having analysed Conradie's work, it is clear to me that his theology is a theology of hope, rather than a theology of optimism.[81] The distinction is critical. He *is* concerned for the redemptive healing of history and that the loving and healing presence of God will permeate the cosmos in the eschaton, but there is little suggestion of the cosmos arriving at some final *telos*. Those brought up on a diet of Karl Rahner and his theological successors might baulk somewhat at the lack of any development of the notion of active self-transcendence and the progressive, ascending history of the entire cosmos.[82] Though, to be fair to Conradie, it is the mechanism rather than the final goal which distinguishes him from Rahner. Conradie *does* suggest, however, albeit briefly, that the 'whole story of the cosmos, the whole cosmic pilgrimage, would reach its completion'.[83]

6.9 Conclusion

There are a number of contributions that may be gained from Conradie's work in developing an ecological eschatology. His work on what he terms three 'predicaments' is significant in the context of eco-theology. The first of these is the predicament of sin. Conradie reminds us of the need for soteriology in the context of eschatology. Based on an understanding of

80. Ernst Conradie, *An Ecological Christian Anthropology: At Home on Earth?* (Aldershot, Hants, England/ Burlington, Vermont: Ashgate, 2005), 229.
81. See Conradie, *Hope for the earth*, 314. Conradie writes that, 'Optimism suggests change will occur naturally and that radical decisions are less urgent'. On the other hand he comments that, 'Hope is often born within desperate situations. It calls for resistance and defiance precisely within such contexts—even when no escape appears possible. To hope is to raise children even when disaster looms.'
82. See Karl Rahner, 'Christology within an Evolutionary View of the World', in *Foundations of Christian Faith: An Introduction to the Idea of Christianity* (New York: Crossroad, 1978), 178–203.
83. Conradie, 'Resurrection, Finitude, and Ecology', 294.

God's complete faithfulness, he prevents any false optimism in the processes of the world by arguing that any concept of self-transcendence or cosmic transcendence must acknowledge the suffering, pain and destruction occurring around us.

The second predicament, to which Conradie alerts us, is finitude in time. Everything is transitory, including the universe itself. The various species rapidly becoming extinct are a reminder of this. Death and transitoriness are not issues which can be ignored in eschatological discussions. Given the predicament of finitude in time, salvation must come from God and be eternal. Conradie's work is particularly helpful in highlighting the strengths and weaknesses of those theologies which explore signs of promise in an evolving universe (especially John Haught's work). Conradie helpfully reminds us that finitude and death are not avoided by a positive understanding of chaos and complexity, including the theory that emerging complexity and top down causation are the means whereby a genuinely novel future may be possible. Conradie provides a welcome balance in reiterating that it is God's faithfulness that counts. Given God's faithfulness, the finitude of the earth (finitude in time) should not pose a problem to eschatology in the end.

The third predicament is that of limitedness in space. Yet God's presence in eternity transcends the cosmic dimension of space (and time). And eschatology ultimately depends on God's transcendence and must therefore be theocentric.

Conradie's two other significant and helpful contributions to an ecological eschatology are his development of the 'two worlds' theory. His exploration of van den Brom's work on the 'depth' dimension gives us a way of articulating an understanding of God's eternity, describing it as a 'dimensional notion of God's transcendence'.[84] The eschaton, then, is the transformation of the cosmos into the heavenly realm.

This notion of 'two worlds' is undergirded by his ideas about 'material inscription'. Here his contribution to ecological eschatology is important. Caring for the earth becomes crucial, because the earth, and all of the various events and experiences that occur on earth, provide the building blocks of the new creation. Yet even those experiences of failure (including sin) can be redeemed, because redemption occurs in God's presence.

For all its strengths, Conradie's eschatology doesn't quite give us all we need for an ecological eschatology, because the subjective, participa-

84. Conradie, *Hope for the earth*, 358.

tory side of the model is left relatively undeveloped. The two very similar two-realm models presented by Conradie and Polkinghorne have helped frame a wider problem in eschatology. Is cosmic teleology a fundamental part of eschatology? The crux of the matter ultimately lies in the definitions used. Both Conradie and Polkinghorne have argued persuasively, through the very presentation of their two-realm models, that the ultimate transformation of the cosmos does not depend on a sense of cosmic completion in the three-dimensional world in which we live. There is a sense of discontinuity in this world. The universe may well come to an end, and indeed the mainstream scientific projection is that, in one way or another, intelligent life of any description in the universe is almost certainly ultimately doomed. If that is the case, then eschatology needs a model which does exactly what Ernst Conradie wishes to do: allow for the healing of the past (as well as the present and the future which will be), so that it may be gathered up into God. In this understanding the ultimate *telos* of the cosmos is to be gathered into God, though this may not happen in history, that is, in the time and space of the created universe.

A criticism may be made of the two-realm model of eschatology: it has more than a hint of neo-Platonism about it. Certainly both Polkinghorne and Conradie are at pains to emphasise the absolutely crucial role and place of the created cosmos in the eschaton. Nevertheless the model lends itself to similar criticism that has been levelled at Augustine's understanding of two cities. Western Christianity will quite possibly always struggle to escape the theological models Augustine developed.

The two-realm model is not identical with the theology of the Augustinian tradition. However, there is still that sense of escaping this world into the next, even though the next world is a transformed version of this world, which sits alongside this world. Be that is it may be, the multi-dimensions (or two-realms) model of eschatology is one challenging and persuasive way of comprehending how this world might be transformed by God: the objective side of a model of ecological eschatology.

The only question is whether the model allows for the sufficient *active* participation in the transformation of the cosmos by human beings and indeed the processes of the cosmos itself. The main weakness of Conradie's work is that it falls short of emphasising the contributory role that human beings and creation have in the transformation of all things. Thus the subjective side of the model of ecological eschatology is not developed by Conradie. The *active* participation of all of creation, both that of human beings as well as the processes of creation, is needed in developing an eschatology which is truly ecological. As I indicated above, the next section will examine the work of Elizabeth Johnson and Alexander Schmemann and discuss how their approach to eschatology might indeed provide a model for understanding the active participation of all of creation in God's transformation of the cosmos *into* the eschaton. I will use their work to develop Conradie's idea of the 'material inscription' of the cosmos, but from a more participatory perspective. Thus this final section will provide us with the 'subjective' side of the model of eschatological eschatology, namely, how human beings and the rest of creation might participate in God's consummation of all things through the process of 'material inscription'.

Section IV

Remembrance, Liturgy and Symbolism

Chapter Seven
Elizabeth Johnson

7.1 Introduction

One of the relevant questions for eschatology, particularly eschatology which is striving to be ecologically sensitive, is that of human participation in the transformation of all things. Anthony Lowes reminds us that, 'in exploring the transformation of the universe, what does not seem to have been explored with sufficient emphasis and clarity is the role of the human person in contributing to this transformation'.[1] Is the work of transformation simply done by 'God, the Trinity, in and through the risen Christ, in and through the Spirit'?[2] Are human beings simply the passive recipients of the transformation wrought by God? This chapter will develop Ernst Conradie's concept of 'material inscription' by looking at Elizabeth Johnson's understanding of memory and its place in eschatology. The question that will be asked is whether in Johnson's work we have a model of eschatology which is both ecological and allows for the participation of human beings, and indeed other parts of creation, in the eschatological transformation of all things. This is the subjective side of the model of ecological eschatology.

Examining Johnson's work from an eschatological perspective may appear to be slightly counter-intuitive, in the sense that her work *prima facie* has a focus on justice, particularly from the perspective of feminism and creation, rather than on eschatology *per se*. However, I think that her theology, particularly her work on memory, can contribute to a model of ecological eschatology. In this chapter I intend to show how we might interpret her work in this way.

1. Anthony Lowes, 'Up Close and Personal: In the End, Matter Matters', in *Earth Revealing—Earth Healing: Ecology and Christian Theology* edited by Denis Edwards (Collegeville, Minnesota: The Liturgical Press, 2001), 136.
2. Lowes, 'Up Close and Personal: In the End, Matter Matters', 136.

In examining the writing of Elizabeth Johnson, one of the critical questions I want to ask of her theology in general and her eschatology in particular is whether it can provide a model for imagining *how* human beings are called to participate in God's inscribing the cosmos into God's life, that is, into the eschaton. This is particularly important in the context of the previous two chapters, which examined the eschatologies of John Polkinghorne and Ernst Conradie. It is crucial that human beings and indeed the whole creation actively participate in this transformation. Otherwise, however convincing the eschatological model might be, the creation and its processes are not valued nor enjoy unqualified integrity. Johnson herself identifies the question of ethics as an issue which must be addressed if eschatology is to have respect and indeed relevance in a post-modern world.[3] Any eschatology must be socially just and ecologically sound.

Johnson's eschatology is based on her development of memory as a powerful tool within the Christian experience of life and particularly the symbol of the communion of saints. The key to understanding her contribution to the field of eschatology begins, then, with an examination of her work on the communion of saints. She describes the whole Christian community as being a part of the communion of saints, or, to use her favoured expression, friends of God and prophets. Her interpretation of the symbol includes vital links between the living and the dead, and it is here that her work on memory, and by extension eschatology, becomes so crucial. Moreover, her horizon of interpretation is not limited to the human race, but 'expands beyond the human world to include the natural world in the community of what is holy and blessed by the Spirit of God'.[4] Johnson's eschatological vision is therefore profoundly ecological, which follows her previous development of ecological themes in other work, as I shall show. Her discussion of what she terms 'the vector of the relationship between the living and those who have died'[5] allows the development of a connection between the two worlds through the category of memory. It is precisely the category of memory which will then, I propose, provide a model for imagining how human beings might participate in the cosmos being 'inscribed' in the eschaton.

---

3. Elizabeth Johnson, *Friends of God and Prophets: A Feminist Theological Reading of the Communion of Saints* (New York: Continuum, 1998), 185.
4. Johnson, *Friends of God and Prophets*, 141.
5. Johnson, *Friends of God and Prophets*, 141.

7.2 Four examples of women's practices of memory

To begin with Johnson wants to develop what she terms four examples of women's practices of memory, as a way of making sense of the communion of saints and providing the link between the living and the dead. Critically, for the purposes of this book, the practices also provide a framework for helping address some of the eschatological problems I have been outlining in this work. The four examples are: 'recovering lost memory', 'rectifying distortion', 'reassessing value', and 're-speaking the silence'.

One of the strengths of Johnson's work is the connection she makes between theological concepts and the stories of the saints themselves. Using examples from the lives of the saints, she brings to life the concepts she is exploring. The first 'friend of God and prophet' to benefit from this methodology, under the practice of 'recovering memory', is Hagar. Johnson describes the story of Hagar as long having been overlooked in the history of Christian teaching and preaching, in particular, because 'it does not fit neatly into themes of divine liberation and covenant'.[6] It is not surprising, Johnson suggests, that the memory of Hagar has been recovered in more recent times by those interested in the perspective of the poor and the outsiders, even those women who suffer from oppression by other women. It is a far different text when read from that kind of perspective, than Paul's use of the story as an allegory in the letter to the Galatians (Gal 4:31), as the basis of a theological argument against circumcision in the Christian practice of faith. Read from the perspective of women suffering from social and economic oppression, the story of Hagar's enslavement (far from her own people and culture), her mistreatment at the hands of her mistress, her life-threatening journey in the wilderness and her incredible will to survive, becomes an interruption to the dominant stories of our culture. 'It demands that the corporate memory of the *ekklesia* make room for the female, the foreigner, the one in servitude, the religious stranger—and the person who is all four—as a vital player in the history of humanity with God.'[7]

This interruption of the dominant stories of our culture remains a theme in Johnson's work, and it is ultimately a theme of hope. However, the story of Hagar herself, as an example of a lost memory, is a troublesome one. It describes the difficult and fraught relations between women in a patriarchal society, in which the begetting of sons was all important. It

6. Johnson, *Friends of God and Prophets*, 142.
7. Johnson, *Friends of God and Prophets*, 145.

describes the economic and social distinctions made and the racial prejudices so abundantly visible. Even God's participation in the story is, if not ambivalent, then difficult. God does not rescue Hagar and her son Ishmael out of the wilderness; she is not liberated out of her harsh environment, in the way that the people of Israel were liberated from oppression in Egypt and, later, Babylon. God gives her a vision to see the resources available to ensure her survival in the wilderness; she sees a well of water. But she is not rescued from her oppressive situation, rather simply given the means firstly to survive and then to create a future. God gives her the capacity to create a better life, even whilst the dire circumstances continue. Different women from different current cultural backgrounds will tell the story with a different emphasis. Thus African-Americans will emphasise her slavery and identify with that, Latin Americans her poverty and servitude and 'the struggle she mounts as a single mother abandoned by her husband'.[8] Asian women on the other hand, reflecting their increasing presence in the global community as servant women working far from home, might emphasise Hagar's loss of cultural identity, whilst African women the social structure of polygamy and the subordination of Hagar to Sarah in the shadow of the great man Abraham.

The recovery of Hagar's story, who is after all 'an extraordinary figure in the history of faith'[9] calls men and women everywhere, particularly those of us in positions of privilege, to a conversion in heart and mind. There is no neat solution to the story of Hagar. She is not rescued from the desert. She does, however, symbolise the struggle of many women (and men) in places of oppression and suffering who, nevertheless, seek a life with God. She makes a way where there is no way. She is an example of a true friend of God: she hears an annunciation about the conception of her son, she receives a divine promise of descendants and dares to name the deity. To this list I would add something Johnson has overlooked: she lives to see her son survive, grow and prosper. She lives to see God's promises begin to be fulfilled in her son: surely of all people, those who are truly blessed are those who see their children begin to show forth the promises of God.

The rediscovery of Hagar's story, lost because of its marginalisation (the prevailing voices of culture making her story virtually invisible), means the rediscovery of a story of resistance and lament, as well as strength and

8. Johnson, *Friends of God and Prophets*, 144.
9. Johnson, *Friends of God and Prophets*, 145. Whether Hagar was an historical figure is a debatable point. Nevertheless that debate does not detract from her significance in the Abrahamic story.

inspiration. Such rediscoveries become an inspiration for action in the present, not only for those who might identify closely with Hagar's story, because of their own suffering, but also for those in positions of power or privilege. The simple realisation that such stories exist can be a powerful agent for change, even to the extent of influencing decisions by those in positions of leadership. Clearly this has significance not only for the lives of human beings, but also for establishing justice in the wider creation.

The second practice of memory which Johnson highlights has an equally restorative function. It is the practice of 'rectifying the distorted story'. The prime example here is the story of Mary Magdalene (Miriam of Magdala, as Johnson reminds us). The distortion of the story has cheated her, and indeed all women, out of the dangerous memory of her discipleship. The distortion, as we shall see, has contributed to the oppression of women in the church, serving to deny them positions of leadership, and creating an image of women as the embodiment of 'female sexuality redeemed'.[10] The power of stories and memories should not be underestimated, and the distorted version of the story of Mary Magdalene has helped form a number of different attitudes in the church with regard to the perception of women and their role in the church.

The distorted story is much more familiar than what we might describe as the true one, as the one closest to the truth, insofar as the truth can ever really be known. The picture of Mary Magdalene that has endured and held a position of power within the church is not that of her contribution to Jesus' ministry, her faithfulness during the crucifixion, and her important leadership role as the first apostolic witness to the resurrection. It is not the picture of a woman healed of a mental illness (in biblical use the presence of 'evil spirits' most often indicates the person is ill, perhaps with epilepsy or mental illness) who then joins Jesus' ministry as a faithful friend and disciple. It is rather the picture, as painted through centuries of patriarchal art, literature and preaching, of a repentant sinner, in all likelihood a prostitute, who is forgiven by Jesus for sins of a sexual nature. Johnson notes that this portrayal robs Mary of the integrity of her own life. It also leaves women more generally 'bereft of a key piece of history on which to build resistance to ecclesial male dominance'.[11] After all, she is described as the 'apostle to the apostles', alluding to the fact that it is Mary

10. Johnson, *Friends of God and Prophets*, 146. Of course, it's entirely possible that the oppression of women led to the distortion of the original story, but Johnson's argument still holds.
11. Johnson, *Friends of God and Prophets*, 146.

Magdalene who plays such a pivotal role in the stories of the resurrection appearances and the re-telling of these stories to the rest of the disciples.

The loss of Mary Magdalene as one of the key leaders and interpreters of Jesus' message to the early church is nothing short of extraordinary, and Johnson rightly probes how this could have happened. She suggests that the distortion of the story occurred through male preachers and biblical interpreters conflating the stories of several different women in the New Testament. Luke's story of the unnamed woman, who anoints Jesus' feet, is combined with John's story of the woman taken in adultery, who is saved from stoning, and his story of Mary of Bethany anointing Jesus' feet. These stories are in turn combined with Mark's account of an unnamed woman anointing Jesus' feet and head with costly nard. All these stories were connected with Mary Magdalene. We might note that this trend has continued in more recent times. The stage show *Jesus Christ, Superstar!* depicts Mary Magdalene as a reformed prostitute. In addition the perceptive and aware viewer of Mel Gibson's hit film *The Passion of the Christ* would have noticed that, in one of the otherwise convincing and helpful flashback scenes in the film, Mary Magdalene recalls Jesus saving her from being stoned for the sin of adultery. The distortion continues in Hollywood films of the twenty-first century! Yet there is no biblical story at all which depicts Mary Magdalene either as a prostitute or any other kind of sinner. Neither is there a story of her anointing Jesus and being forgiven her sins. 'Yet these various stories were coalesced under her name, with the result that by medieval times she becomes a sinner repenting her ways in the sex trade.'[12] Johnson suggests that this distortion has been understood by feminist writers as the shaping of religious symbols by men into a patriarchal framework. Whether the distortion is unwitting or not, and Johnson wisely refrains from distracting her reader by entering that particular debate, the picture of Mary Magdalene which emerges is certainly not the creative leader chosen by Jesus. Rather the conflation of those several biblical stories, and the artistic depictions of her as a beautiful, voluptuous woman, with dishevelled clothing, has created a picture of her as an attractive, unattainable object of desire. It has also removed from the popular stories of the church an example of a wise, faithful follower of Christ, who became an apostolic leader, speaking and acting in the power of Spirit-Sophia and continuing to preach Jesus' message once he is gone.

12. Johnson, *Friends of God and Prophets*, 149.

The rectification of this distorted story brings something quite significant back to the church. Obviously feminist strategies of interpretation bring a different and clearly more authentic chronicle to light. But not only is Mary's story, and the memory of Mary, given back the integrity it deserves, it also serves as an inspiration to all who would faithfully follow in Mary's steps. The rediscovery of a more authentic version of the story allows the memory of Mary Magdalene to cheer women in particular with a 'dynamic flash of faithful friendship, spiritual insight, and self-confident leadership'.[13] But for all of us, men and women, it can serve as an inspiration for action. The correction also reminds us that even the most outstanding witness can be warped by the pressures of patriarchal dominance. This reminder of course applies not just to the stories of human beings, but also to those of the natural world. Where the natural world has been abused, and its story distorted in the process, the rediscovery of a more authentic story can promote a profound sense of the need for ecological justice.

The third category of women's practice of memory Johnson describes as 'reassessing value', by which she means reassessing the value given to a particular memory in the church's history. Often, indeed usually, the patriarchal value assigned to a particular memory is not at all the value which others, particularly feminists, would assign. The example of the virgin martyrs in the early church is a good one and well chosen by Johnson. The reassessing of the value of this memory can, like the previous two categories, enhance the doctrine of the communion of saints 'in ways that empower women and challenge the dominant social-political patriarchal world'.[14] The context of this memory is the traditional interpretation by the church of female virginity as a high spiritual value. In the traditional version of the various stories, young Christian women protect their virginity to the point of their intended husbands betraying their Christian commitment to a magistrate, thus bringing about their death. The memory is preserved and worthy of remembrance because these young women defended their virginity and suffered for it, thus underscoring the value of virginity and suffering as a means to perfection for women. The darker side of the memory, of course, is that women who in contrast exercise and enjoy their sexuality are perceived as being less spiritually virtuous.

But the feminist approach to the memory of these martyrs points us in quite a different direction. Marriage and the demand to reproduce were

13. Johnson, *Friends of God and Prophets*, 150.
14. Johnson, *Friends of God and Prophets*, 151.

tools that the Roman society used to replenish the far flung legions of the empire. So, in the context of Roman history, the Roman state was not only exercising control over women and their fertility (like all patriarchal cultures) it was also using women for its political purposes. Given the true cultural background to the martyrdom stories, we see quite a different memory emerge in place of that of the pious Christian virgins, whose example all women should emulate. We see rather the defiance of these young women in the face of an all-powerful patriarchal culture and their decision to define their own identity: 'They take control of their own life. How fierce a young girl's sense of honor can be! How this has always been feared and repudiated by male authorities.'[15] The stories, thus reassessed, reveal quite different characteristics of the women from the ones we normally associate with the virgin martyrs. The stories show us women, who define their own identity, and who are committed to God as the ground of their identity and wholeness, even if their very lives are threatened. These stories can be read in a way which then strengthens the vitality of women's self-definition. These women, attracted by the mystery of God, 'seized the opportunity to be protagonists in their own history. And in that moment they meant something dear for the hope and self-understanding of their fellow Christians.'[16]

Crucially, however, Johnson's interpretation of such memories doesn't end with a feminist spiel about self-identity. Rather she offers the suggestion that these stories, despite the fact that they ended in death, might nevertheless become the inspiration and hope that energises continuing resistance on the part of others. The terrible death experienced by the virgin martyrs (and indeed other Christian martyrs) is, however, not the only road to salvation. Survival is also a holy choice, Johnson suggests, for in surviving women may also define their own identity, inspire others, and deepen the possibility of future healing and rebuilding. Thus resisting death, especially if it means bringing life to others in the process (such as dependent children in South America, the locus for many of Johnson's examples), can also be a way of being holy in the world. Thus the reassessed memory of the virgin martyrs takes its place amongst the many women who courageously and compassionately fight for survival for themselves and others.

The final category Johnson explores here is that of 'reclaiming the silence'. As the title implies this practice of memory involves bringing to

15. Johnson, *Friends of God and Prophets*, 153.
16. Johnson, *Friends of God and Prophets*, 155.

mind all those women whom history has forgotten, those who are now invisible and forgotten. Remembering these women may not mean remembering outstanding women. It may simply mean connecting, through the very act of remembering, with women of every age who struggled along, usually in patriarchal worlds, 'finding connections to the sacred in the midst of myriad daily sufferings and joys'.[17] Remembered too are the countless victims of war, violence and oppression. So we remember the witches persecuted by the Inquisition. We remember the comfort women supplied to the Japanese soldiers during the 2nd World War and the enslavement of African women by their new-world, Christian masters. Choosing to remember these unknown women, releases strength in the struggle for an unknown future. It 'gives rise to a surge of awareness, grief, gratitude and hope'.[18] Inspired by their memory we continue our struggle for justice now and in the future. There is an obvious connection here with the silent voice of creation, which all too often has been catastrophically abused. Yet its non-verbal cries of pain and anguish have not been heard by its destroyers. To reclaim the silence of creation is also to be inspired to ecological action in the future.

Interestingly, Johnson's work here doesn't include the notion of remembering the dead before God *in order* to commend them *to* God. It is a curious omission, given her Roman Catholic background. As we shall see, the closest she comes to this notion is the belief that 'we dare to trust that Holy Wisdom has not allowed them to perish but has received them into unimaginable life forever'.[19] But the understanding that in remembering the dead we also commend them to God has a great deal of merit to it, provided that we steer away from the idea that such prayers ensure their eternal salvation.[20]

All of the above categories of memory are used by Johnson to revitalise the church's doctrine of the Communion of Saints. Those to whom the church in its patriarchal construct had assigned a particular value, or

17. Johnson, *Friends of God and Prophets*, 157.
18. Johnson, *Friends of God and Prophets*, 159.
19. Johnson, *Friends of God and Prophets*, 251.
20. Certainly we need to be careful of returning to a variation of the mediaeval concept of indulgences. The idea that a good deed, or the purchase of an indulgence, might shorten a loved one's stay in purgatory, is certainly to be rejected at all costs. But just as we pray for the living and commend them to God's care, so too the doctrine of the communion of saints, understood in the broadest possible way as Johnson does, might include that sense of praying for the deceased and commending them to God's merciful love and protection.

whose memory the church had simply forgotten, distorted, or never preserved at all, are brought back to life by the practices of women's memory. From an eschatological point of view, this use of memory can describe both God's role in recalling past events to his mind, and *our* role in bringing them to God in acts of liturgy and worship. Importantly, following on from Johnson's notion that the communion of saints is not confined to human beings but includes the whole of creation, these practices of memory can also be used to recall parts of creation itself. A simple example will begin to illustrate the methodology I am proposing, though it is a concept I discuss further below. The example of the practice of reassessing the value of a memory can quite easily be applied to creation more generally. The clearing of land in parts of South East Asia and South America for the purposes of the commercial grazing of cattle is a good example. The *value* assigned by the dominant consumerist and commercial forces of our culture to that action, and hence ultimately to the memory of that action, might be quite different from that assigned by individuals concerned with ecology. The dominant capitalist forces, the ones indeed responsible for clearing rainforests and the like, assign a commercial value to the land, based on the number of cattle it can support and so on. The memory of the action is therefore formed by the capital value of the land.

But for individuals and societies concerned about ecology the memory will be quite different. It will be one of suffering and loss: the loss of a natural habitat, the suffering of creation itself in the process and the ongoing suffering of a land abused by cattle never meant to be there. When this memory is reassessed, or reclaimed, it can be brought before God in liturgy (the liturgy can take any number of forms). This is one way we participate in God's inscribing the cosmos into God's life, that is, into the eschaton. But the memory brought to God is that of suffering and pain, as well as the memory of the creation *before* its exploitation and oppression.

But there is another, perhaps more practical and less eschatological, consequence of this practice of memory as it applies to creation. It provides us with the inspiration for resistance and action in the present. Thus memory becomes not just an eschatological act, in which creation is brought before God, that God will materially inscribe it into God's life—it is also the basis for an ecological ethic. It provides the ground for ecological action, where necessary. The memory of the past becomes the motivation for action in the future. The memory of the creation we have hurt or destroyed creates within us a desire for orthopraxis.

The category of reclaiming the silence has a place in the bringing of all of creation before God too. Those countless examples of violence done against nature, too numerous to recall in details, many of which are now lost to memory, are nevertheless brought before God. Just as in the category of 'reclaiming the silence' we remember, for example, the tens of thousands of women persecuted in the witch hunts, so too the many acts of ecological degradation, too many to name, even if we could recall them all, are brought to God.

7.3 Companions in memory

Johnson's eschatology is built on an understanding of memory and its significance within the Christian tradition, particularly when remembering the forgotten saints from a feminist perspective. Whilst this feminist perspective is not used by Johnson in an exclusive way, that is, in a manner which prevents memory from being used in broader categories, nevertheless it is critically important in her work and provides the place to begin my examination. Johnson describes memory in the feminist context as searching for and remembering the 'lost heritage of women's tough and joyful lives in relation to the mystery of God'.[21] The participation by women in such exercise of memory not only helps restore 'a heritage of religious power to a disenfranchised group'[22] but also helps the women themselves gain an appreciation of their own blessedness and thus become energised through this spiritual event 'to bless, to work for justice, to follow Christ by forging new paths of discipleship'.[23] Not only are women thus empowered to play critical roles in establishing justice, but also, by recognising and embracing their own sacredness, women are empowered to name and celebrate the lives of other holy women. The saints of today (as Johnson terms these women) search out forgotten saints and bring them to the attention of the wider church community as symbols of the sacred.

This dynamic interaction between the saints of today and the saints of the past, made a dynamic symbol and not a static doctrine, thanks to the vivifying work of the Spirit, gives energy for the struggle now and points to a different future. Exactly because women in particular search out these

21. Johnson, *Friends of God and Prophets*, 163.
22. Johnson, *Friends of God and Prophets*, 163.
23. Johnson, *Friends of God and Prophets*, 163.

past saints, they are given inspiration and energy for the struggle for justice now.

In describing this process of remembering the forgotten saints of the past Johnson uses the template 'narrative memory in solidarity'. Johann Baptist Metz, one of the principal shapers of this template, suggests these are basic categories of a theology 'that seeks to interpret the love of God in the midst of a conflictual world'.[24] The three categories, 'subversive memory', 'critical narrative' and 'solidarity in difference', help prevent theology becoming too abstract and remotely academic, whilst a suffering world continues on. They prevent this by enabling a 'grasp of faith inseparable from active discipleship, so that the community of memory and hope becomes through its prayer a community that struggles for justice, peace and the well-being of the earth'.[25]

The first category, 'subversive memory', is a significant one for Johnson in her attempt both to ensure that the memory of forgotten saints is not lost forever and also to enable the memory of the saints to play a continuing, pro-active role in the transformation of contemporary society. She suggests that amnesia robs a person of an irreplaceable sense of who she or he is. A dominating, oppressive regime will attempt to keep a group submissive, by destroying the memory of its heritage, ancestors, and traditions. But subversive memory, which dares to connect with loss and 'the pain, the beauty, the defeat, the victory of love and freedom, and the unfinished agenda of those who went before, acts like an incalculable visitation from the past that energises persons. It interrupts the omnipotence of the present moment with the dream, however fleeting, that something else might be possible.'[26]

By remembering sufferings and victories of the past those who are bored or despondent are startled into movement. In times of banal plenty, it questions whether there might be more to life than the 'acquisition of things and the search for the latest entertainment'.[27] In our contemporary, developed, 'Western' nations this being startled into movement is itself a life-giving event. There is almost an apathetic stupor which can descend upon societies fat with the spoils of their wealth. Ultimately this stupor is destructive and one needs only examine the pages of history to discover other nations and empires which, once great, began a slow but inevitable

24. Johnson, *Friends of God and Prophets*, 164.
25. Johnson, *Friends of God and Prophets*, 164.
26. Johnson, *Friends of God and Prophets*, 165.
27. Johnson, *Friends of God and Prophets*, 165.

decline as they enjoyed the wealth of their military and economic successes.

On the other hand in difficult times it calls oppression into question saying, 'There are many who have struggled before you and it is within you to resist the apparently inevitable course of events'. In societies and situations of oppression this kind of remembrance is particularly hopeful. In remembering those women and men who resisted violence and oppression in their own day, women and men today are given courage and inspiration to do just that themselves in the present.

This kind of remembrance was called by Metz 'eschatological remembrance' and Johnson attaches great importance to it, because there is a surplus of meaning carried in this act of remembering, which makes it possible to create a different future.

Johnson rightly points out that such practices of memory are potentially dangerous, exactly because the memory brings to light new possibilities and empowers hope for a different future by detaching people from the grip of a grim situation, from what is currently held to be plausible. Johnson suggests that the classic example of this kind of remembrance is the memory of the passion, death and resurrection of Jesus Christ. The memory of suffering, injustice, violence and death, and through this God's victory over evil and death, 'grounds the promise of future freedom for all the world'.[28] This memory, with its hope for a promised but unknown future, sustains our Christian community's efforts to live in the world faithfully and compassionately.

The central point of Johnson's argument here is that this memory challenges the triumphant pride of modern developed societies and its callous, bourgeois complacency. Importance and significance are given to those defined by the dominant and powerful as unimportant (including all those historical victims vanquished by injustice), 'while hope in the resurrection anticipates a liberating future precisely for them'.[29] Here Johnson's concern and passion for justice becomes abundantly clear. She suggests that the capacity to remember jolts complacent, apathetic Western societies into action, whilst giving power to the oppressed in all societies. Moreover, the *ekklesia* is propelled out of passivity into active engagement against all that oppress the glory of God in the world.

Johnson seeks to revitalise the understanding of the communion of saints by emphasising this subversive, resistant, compassionate under-

28. Johnson, *Friends of God and Prophets*, 168.
29. Johnson, *Friends of God and Prophets*, 168.

standing of it. Including and remembering the whole cloud of female witnesses serves to underline this subversive character. The challenge is to re-shape the church's memory, so that women are included and the whole memory reinvigorated as a result. Johnson's point about the letter to the Hebrews is well taken: the cloud of witnesses referred to in Hebrews 12:1 helps the runners; they are sustained by the witnesses, not hindered. The memory of female saints, in the context of Johnson's feminist approach, and that of their often unfinished agenda, is an inspiration to bolster our own efforts. Indeed through our memory the cloud of witnesses are present and drive us to action.

After the category of 'subversive memory', the second category which Johnson defines is that of 'critical narrative'. The transformative power of a story is made manifest through the dynamic of the narrative itself: through the inter-relations between the story, the story-teller and the listeners. It has the power to change its listeners or to lure them into certain attitudes and behaviours. The martyr stories of popular Catholicism in El Salvador in the 1980s are a powerful example of the power of narrative. The story of Jesus was not simply told, but lived in. 'Battered by poverty and struggle, the death of a beloved archbishop, the violent deaths of loved ones, and ongoing suffering, people found in the paradigmatic story of Jesus' life, death, and resurrection both an interpretation and a spur to action.'[30] Thus the story of Jesus becomes a summons to emulate Jesus' actions. The memory of Jesus, told as critical narrative, urges and inspires individuals to follow in his steps and take action.

In this example of El Salvador we see also the dangerous side of memory. The stories about Jesus, when understood as South American stories, in other words stories which were extraordinarily pertinent to the South American situation, became dangerous. They were dangerous precisely because they prevented apathy on the one hand and despair or despondency on the other. They allowed people to resist manipulation, to prevent the loss of heritage, culture and customs, and to inspire transformation and renewal. Narrative finds voice when fear silences other forms of communication. Narrative finds voice when reason and logic are silenced in the face of suffering. 'In the classic tradition the experience of suffering generates an awareness of the apophatic dimension of faith: there are no reasons and silence becomes a suitable response. Refusing to trivialise

30. Johnson, *Friends of God and Prophets*, 171.

tragedy, narrative finds a home in this dimension, birthing at least glimmers of light in the midst of darkness.'[31]

The Christian story of Jesus of Nazareth is a powerful example of the narrative genre being used to allow communal behaviour and personal identity to flourish and political justice and meaningfulness in all areas of life to exist. All aspects of the narrative about Jesus speak powerfully to situations of oppression, suffering and violence. His table-fellowship with 'sinners', tax-collectors and prostitutes, his openness to and love of the poor and the outcast, his confrontation with the self-righteous and hypocrites, the parables and Jesus' own, tremendous story-telling ability, all speak to people in situations of fear. And they are all set against an eschatological horizon. It is only then that the promise of the narrative will be fulfilled: when the living and the dead and indeed the whole of creation are consummated in God. Until then the narrative is dangerous and subversive. It gives life and hope to the Christian community. Telling the narrative means oppression and violence, and the history written by the victors, is not granted legitimacy. This is particularly true when we remember martyrs—the name means witness—because they witness to theology, and to us, a different interpretation of events:

> Thus, remembering the dead, telling their stories, making lists, creating art, reciting martyrologies, keeping memorial days, tolling bells, crafting litanies, all strike a blow against oppression. Narrative memory refuses to grant it legitimacy.[32]

Told in such a way, stories using the critical narrative method refuse to let the stories of the victims and the marginalised to be oppressed and manipulated by the dominant forces of society. The stories are told in defiance of the oppressive patriarchal powers of the world, lending strength to the ongoing task of transformation: 'The critical narrative of the cloud of witnesses functions then as a verbal sacrament of the vivifying, redeeming work of the Spirit, helping women stay afloat and issuing a challenge for transformation.'[33]

Finally we come to Johnson's third category of remembrance, which is 'solidarity in difference'. When we remember and identify with others

31. Johnson, *Friends of God and Prophets*, 173.
32. Johnson, *Friends of God and Prophets*, 174.
33. Johnson, *Friends of God and Prophets*, 175.

with the kind of solidarity she is advocating, our connection with them becomes so strong that their joys and pain become ours and a spur to transformative action. This communion includes the living, in which case solidarity may mean joining initiatives of struggle or, when not directly victimised, by choosing to love by taking the pain as one's own and struggling for life for all. But it may also include the dead, especially those marginalised, overcome or defeated. This solidarity backwards through time emphasises 'the common character of human destiny and its unfulfilled promise'.[34]

Solidarity in difference has been lived out by God in the person of Jesus Christ. Nowhere is God's true intent of salvation and ultimate well-being for all more powerfully expressed. Through God's commitment to incarnate solidarity with the world, made manifest in the risen Jesus' life and death and the ongoing presence of the Holy Spirit, we see the kind of solidarity demanded of us in return. When we participate in solidarity, then we can remember in the sense being described by Johnson.

It is only openness of heart, and intense listening and imagination, which allows this solidarity to be accomplished across boundaries and despite differences. In a moment, almost of confession, Johnson acknowledges the criticisms made of white, educated, middle-class women, who tend to overlook differences among women themselves and project their own experiences as universal. Responding to these concerns feminist theologies strive to be accountable to differences among women's situations and practical concerns world-wide. But these differences themselves promote creative action in the Spirit, through this sense of connectedness. The diversity down through history, where women and indeed all the saints, achieved different, even apparently strange and alien things, gives hope and courage to us now.

When the template 'narrative memory in solidarity' is used, it gives life to the doctrine of the communion of saints. It becomes practised less 'as a cult of the saints and more as a discipline or way of remembering that brings life. Then the communion of saints becomes the cloud of witnesses, surrounding those who cry for justice with encouragement and blessing, lending the support of their own witness for personal and social transformation.'[35] This template allows us to hope, albeit in the apophatic tradition, that we too will join the communion of saints. It gives us the

34. Johnson, *Friends of God and Prophets*, 176.
35. Johnson, *Friends of God and Prophets*, 180.

impetus to struggle for justice now, thus fulfilling one of Johnson's crucial criteria, that justice is an integral part of eschatology.

Her concern for creation, that eschatology should have an ecological dimension, finds expression in her insistence that the communion of saints includes the whole of creation.[36] This is directly relevant to eschatology, as I will show below.

7.4 Problems with traditional eschatology

Johnson's work is comprehensive, because she does outline many of the problems associated with traditional eschatological concepts and their frame-works. One of the fascinating issues, from my point of view, that she comes to terms with, is the question of how the saints of earlier generations might still be present to us. She poses the problem with the following question: 'Do the dead live on only in the memory of the living or, in some unimaginable way, do they themselves also live anew by the power of the vivifying Spirit of God?'[37] Fundamental to her understanding of the influence the saints might have in our lives is a rejoicing in the belief that the saints are 'enfolded into the absolute mystery of the gracious, compassionate being of God which to us is darkness but which to them is the fulfillment of their lives in the sphere of the Spirit. Their destiny signals the future toward which we the living are headed.'[38]

In our contemporary, post-modern culture there are a number of issues Johnson must deal with in order to maintain an understanding of the dead being alive in the Spirit. The medieval synthesis of the doctrine of the last four things, death, judgement, heaven and hell (with purgatory being an intermediate state of purification leading to the enjoyment of the beatific vision of heaven) has met with its share of criticisms. Johnson categorises the criticisms into two main sets. Firstly, there are intellectual problems with the concept of life after death: recent and contemporary philosophy has rejected the notion of the duality of the body and the soul and suggests that it is untenable that they could endure separation. Furthermore contemporary science agrees with philosophy in arguing that matter and spirit are not fundamentally different substances. Even though once spirit emerges it is not simply reducible to the workings of matter,

36. Johnson, *Friends of God and Prophets*, 242.
37. Johnson, *Friends of God and Prophets*, 181.
38. Johnson, *Friends of God and Prophets*, 214.

nevertheless spirit and matter 'are two forms of the same substance'.[39] Interestingly Johnson does not address this particular question. Perhaps she feels it is beyond the bounds of her particular expertise. Certainly the question of the relationship of the spirit to matter is one of the fundamental challenges facing any understanding of the doctrine of resurrection of the dead. This is true not only of the relationship of individuals to their earthly bodies, but also of their relationship to the earth itself.

The other main set of problems, which Johnson identifies, is the question of ethical problems with the concept of life after death. The criticism has been made that 'religious interest in life after death has been so otherworldly that it has robbed the earth of intrinsic value, cut the moral nerve in the face of injustice, and sapped energy needed for the struggle for life over death here and now'.[40] This criticism builds on the Marxist critique that religion is the opium of the people when it comes to social justice. However, it includes a concern that promises of life after death also rob the Christian of his or her concern for the environment. Given the increasing global ecological crisis there is an opinion that instead of focussing on the hereafter, our minds need to be focussed on this earth. In other words Christianity, and eschatology in particular, must become 'green'.

Johnson also identifies the feminist critique of the 'deleterious effects of patriarchal dualism'.[41] She suggests that the image of an immortal soul in a corruptible body disparages women, their female bodies and other feminine functions, which are associated with the 'lower, messy, physical order from which one ascends to the heavenly, masculine realm of spirit'.[42] Whether this dualism has added to the association of women with the lower physical order remains a contentious issue. But certainly the dualist attitude to the body and the soul has been unhelpful in terms of valuing our physical existence on this earth.

7.5 Reinterpreting traditional symbols

Some of Johnson's most interesting work occurs when she looks at some of the attempts to reinterpret the traditional symbols in the light of these contemporary criticisms. The understanding of purgatory has certainly developed in critical theological thinking. It is possible to think of human

39. Johnson, *Friends of God and Prophets*, 184.
40. Johnson, *Friends of God and Prophets*, 185–186.
41. Johnson, *Friends of God and Prophets*, 185–186.
42. Johnson, *Friends of God and Prophets*, 185–186.

development not being frozen at death; the concept of purification happening 'instantly' after death, rather than through a process of 'time' has also become more popular. Many contemporary reinterpretations of the doctrine have much to commend them. I would agree with the assertion that the consequences of our actions in life 'must be dealt with as part of coming into communion with God after death'.[43]

The reinterpretation of the doctrine of heaven is one which is of interest to all Christians. I think Johnson is absolutely correct when she asserts that: 'At root, heaven is the symbol of a community of love sharing the life of God'.[44] Traditional images of heaven have the danger of being decidedly more boring than those of hell! Johnson writes that we must talk about knowing God 'in the biblical, experiential sense, relating intimately to the unfathomable mystery of another in deeply mutual regard'.[45] Any development of our understanding of heaven needs to occur along these lines. Unfortunately the traditional symbol of heaven as the beatific vision has suffered somewhat in the telling.

Many theologians have a number of concerns about the symbol of hell. It seems to contradict contemporary thoughts on God, because of our understanding of God as the God of love, who would never wish to damn one of his creatures. However, it is possible to come to an understanding of hell if we see hell as a choice persons make when 'they opt to shut love for God and neighbour out of their lives and pursue a genuinely evil course'.[46] Johnson points out that other theologians, for example Hans Urs von Balthasar, argue that we are permitted to *hope* for the possibility of universal salvation.[47] It is indeed interesting to note that whilst the Roman Catholic church has canonised people, it has never presumed to pronounce that particular individuals have indeed been damned. I share some sympathy with von Balthasar's opinion and wonder whether, faced with the overwhelming love of God in death, anyone could actually choose to reject it.

Certainly the process of critical reinterpretation has its problems and Johnson asks whether we can rescue the symbols for use. Here I begin to appreciate the mastery of Johnson's writing. She allows for mystery and

43. Johnson, *Friends of God and Prophets*, 188.
44. Johnson, *Friends of God and Prophets*, 190.
45. Johnson, *Friends of God and Prophets*, 190.
46. Johnson, *Friends of God and Prophets*, 191.
47. Johnson, *Friends of God and Prophets*, 181. See Hans Urs von Balthasar, *Dare We Hope 'That All Men Be Saved'?* (San Francisco: Ignatius Press, 1988).

darkness when discussing these issues, whilst not allowing that sense of mystery to be an excuse not to discuss difficult questions. She is at her best when she comments on the power that these symbols can potentially have in people's lives. It is this concern of hers for the faith of the people which keeps her writing fresh and relevant. Whilst commenting on the problem of reinterpretation she writes:

> Realising that at bottom these symbols signal the hope at the heart of Christian faith and alert people to the seriousness of how they take up their life here and now may still allow them to function with some suppleness in the community's language of faith. Even when reinterpreted, the traditional imagery and its explication in theology now create a sense that the future after death is more open, more of a mystery, than we had imagined.[48]

Johnson is correct in adopting an attitude that the future after death is unknown and more of a mystery than we have traditionally allowed for.

Some of Johnson's most incisive work occurs when she goes on to examine what she terms 'Narrative Scenarios', which are one form of response to traditional eschatology. She examines in some detail Rosemary Radford Ruether, who considers that death should be accepted as a natural part of the life cycle. Traditional eschatology distracts us from our task of 'creating a just and peaceable society on this planet to hand on to our children'.[49] She argues 'individual mortality is a symptom of the basically patriarchal attempt to transcend mortality, an egotistic effort to absolutise oneself over against the total community of being'.[50] Instead we should allow ourselves to be reabsorbed into the 'great Matrix of being',[51] because 'being recycled into the universe of being is our ultimate self-gift to the world'.[52]

Johnson applauds Ruether's concern for ecological consciousness, but she suggests that 'a solely immanentist end is not the only conceivable

48. Johnson, *Friends of God and Prophets*, 192.
49. Johnson, *Friends of God and Prophets*, 196.
50. Johnson, *Friends of God and Prophets*, 196.
51. Johnson, *Friends of God and Prophets*, 196.
52. Johnson, *Friends of God and Prophets*, 197.

possibility, even within an ecological and feminist perspective.'[53] Third world feminist liberation theologies suggest:

> that hope for life with God after death for human persons and the whole earth not only does not cut the nerve for action on behalf of justice but actually sustains it, especially in violent situations. Furthermore, such transcendent hope, when cast into a nondualistic framework, functions critically and creatively to promote care for the earth precisely because it sees that this world has eternal value.[54]

Johnson's brief work on third world liberation theology is reflected in the writings of black American theologians.[55] What might be termed transcendent, eschatological hope is by no means inconsistent with a passion for justice and equality now. Indeed, the hope that the victims of history will be vindicated by God seems to serve, somewhat ironically, as an inspiration for a passion for justice now.

So, in my opinion, Johnson has identified the crucial issue in this whole debate when she suggests we need an eschatological goal which is both immanent and transcendent. She then compares the work of Karl Rahner, with his emphasis on transcendental anthropology, with the 'Dissolution Scenario' of Bartholomew Collopy. In Collopy's theology death is seen as being truly destructive and unnatural. Only after wrestling with this nothingness of death will faith emerge in its rawest form. We risk that God will be there, holding on to the image of the cross of Christ. 'In this approach, there is no survival of death that would not be based on the most creative act of God.'[56]

Johnson's own contribution to the debate is quite welcome. She suggests that expressions of the character of God (especially as expressed in God's relationship to the world) can be made particularly in pneumatological terms. She sees the Spirit as the 'creating, redeeming, sanctifying God' who is 'always and everywhere present and active in the world, drawing near and passing by to bring life and resist evil.'[57] The three acts

53. Johnson, *Friends of God and Prophets*, 197.
54. Johnson, *Friends of God and Prophets*, 197.
55. See for example James Cone, *God of the Oppressed* (Maryknoll, NY: Orbis Books, 1997).
56. Johnson, *Friends of God and Prophets*, 200–201.
57. Johnson, *Friends of God and Prophets*, 206.

of the Spirit which suggest that we are destined not for re-absorption but for eternal communion are creation (where the Spirit gives life and draws matter into increasing complexity), the resurrection of Christ (which is a precursor to what will happen on a cosmic scale) and the self-communication of the Spirit in grace. Johnson suggests that this interpretation of divine action suggests that God will continue to act in the same blessed ways in the future.

Thus, returning to the question of the communion of saints, we may view the saints not simply as a company of friends, who are remembered and are inspirations to us in our lives of action, justice and love, but also 'as a company of friends whose destiny shines as a beacon of hope'.[58] Having been transformed, they are our companions in hope. This hope can have critical power: because the poor are indeed promised a place in heaven, they can resist structures which deprive them of that place, even now. Moreover, because the world itself will be transformed and because we have the promise of a God-given future in the 'vertical' sense as well as the 'horizontal' one, what transpires here and now is of lasting importance. In addition, as Sallie McFague suggests, we have a vocation to bond with all creatures to create a rich sustainable world here and now, which is the creative function of hope.[59]

The remembrance of the saints (in the sense of *anamnesis*), particularly through such wonderful liturgies as the recitation of the litany of saints, allows them to be transformative in our own lives. Johnson is correct when she writes that a remembered event (or indeed a remembered person) 'becomes a living force in history when it is recalled and narrated; in the very retelling power comes forth to change the horizon of our days and offer new possibilities of existence'.[60] This is the same process which many Christian understand as occurring in the celebration of the Eucharist, when we celebrate and remember Christ's life, passion, death and resurrection. In remembering those events, they indeed become transformative in our lives.

The strength of Johnson's position is that she acknowledges the problems with traditional eschatological categories, but nevertheless affirms the importance of hope in a setting which is not solely immanent. Her position is also strengthened by her reluctance to spell out too clearly or

58. Johnson, *Friends of God and Prophets*, 214.
59. Johnson, *Friends of God and Prophets*, 218. See Sallie McFague, *The Body of God: An Ecological Theology* (London: SCM, 1993), 202.
60. Johnson, *Friends of God and Prophets*, 234

precisely how this might occur or what shape our future may take. She works from the perspective of the apophatic tradition, which suggests that in one sense we cannot know how even to begin to describe what is traditionally called heaven, let alone what is meant by a 'new heaven and a new earth'. Nevertheless, whilst we are quite correctly reluctant to enunciate clear concepts about life after death, God can be trusted. God is faithful and God's fundamental love for us and for the earth, creation, means that our hope need not be restricted in the way that Rosemary Radford Ruether tends to do.

Johnson has revived my own passion for and interest in the communion of saints. Her work on women's experiences of memory and the recovering of the memory of Hagar, the rectifying of the story of Mary Magdalene and the reassessment of the value of the virgin martyrs helps return these early saints to their rightful, prominent place in the history of Christianity. Her work also honestly but creatively addresses the postmodern criticisms of traditional eschatology. I am left with two images from Johnson's work, which sum up what she attempts to achieve in her critical, reinterpretation of the doctrine of the communion of saints. The first is her summary of John Zizioulas' ideas:

> In other words, since a person is essentially constituted by relation to God, then even in the shattering of death one is held in communion by this relation of Love, which always and everywhere creates a new being. Love is the voice of eternity in our heart.[61]

God loves us so much that even the shattering experience of death does not separate us from him.

The second image is Johnson's description of the passage in the letter to the Hebrews: 'In Hebrews the cloud of witnesses is not a handicap to the runners, weighing them down, but rather a dynamic help that sustains their efforts, helping them run faster'.[62] Johnson's work will help her postmodern readers rejoice in the presence of all the saints and encourage them on their own journeys of faith, striving for justice, peace, the integrity of creation and love.

---

61. Johnson, *Friends of God and Prophets*, 194.
62. Johnson, *Friends of God and Prophets*, 169.

7.6 Ecological connections

From the perspective of ecological eschatology, Johnson's most interesting and helpful line of thought is that the communion of saints does not simply refer to human beings:

> Set within the life-giving history of God with the world, which is not simply an anthropocentric event, the *communio sanctorum* ultimately reaches out to signify the community of all creation, past, present, and to come, sheeting in the flow of life in the Spirit: holy people and a sacred earth together.[63]

Thus, although Johnson's work is actually mostly anthropocentric in focus when she writes about the communion of saints, it is nevertheless readily adapted to include our solidarity with creation itself and remembrance of those parts of creation which have been victimised or destroyed completely. For example, in an Australian context, the remembrance of the destruction of Lake Pedder in Tasmania becomes a spur for environmental action in the future. Our *subversive memory* means that the history written by the victors of that particular disaster is never the final word. Our *critical narrative* can take on the shape of the liturgy previously reserved for the martyrs. The development of 'green' eucharistic thanksgiving prayers is another practical example. Our *solidarity in difference* is our sense of connection with environment; despite our differences with other creatures, both plant and animal, we acknowledge our shared identity as having been created by God.

The link between human beings and all of creation in the communion of saints, and the *memory* of this inclusive understanding of the communion of saints, then becomes a model for ecological eschatology. The model depends on the understanding of human beings and the world being intrinsically connected. Everything and everyone is connected to everything and everyone else. Nothing stands alone. Indeed, as she reminds us, in Moltmann's work all created things interact with one another and with God through the dynamic presence of the divine Spirit. All of creation waits for redemption through Christ who is, 'the firstborn of all creation'[64] (not just human beings).

63. Johnson, *Friends of God and Prophets*, 242.
64. Colossians 1:15.

The final issue to be examined now is how human beings and the rest of creation might participate in the transformation of the cosmos (the process of 'material inscription', to Conradie's term) in a way which includes remembrance, but includes other activity too. The task is to elaborate further on how all of creation might actively participate in the transformation of the cosmos on a day to day basis. When we engage in *solidarity in difference*, to use Johnson's term, what are the specific activities (including remembrance) that are involved? In the next chapter I will examine liturgical theology, particularly the eucharistic theology of Alexander Schmemann, and discuss how this also provides us with a model of active participation. We participate in the process of 'material inscription', when we give back to God the things that are God's, and ask him to bless them. But this process of giving back to God, or praising God, is not one restricted to human beings, but includes the non-human creation too.

7.7 Conclusion

Johnson's work on the category of memory is extremely helpful, because it provides us with a model for imagining how we might participate in the cosmos being 'inscribed' in the eschaton. Her four examples of women's practices of memory, which she gives as a way of making sense of the communion of saints and providing the link between the living and the dead, provide a framework for helping address some of the eschatological problems I have been outlining in this work. The four examples are: recovering lost memory, rectifying distortion, reassessing value and re-speaking the silence. Using Johnson's categories of memory in an ecological context, creation itself can be brought before God in liturgy and thus inscribed into God's memory.

Likewise Johnson's template 'narrative memory in solidarity' can also be used in an ecological context. The category of subversive memory brings to light new possibilities and empowers hope for a different future by detaching people from the grip of a grim situation. The category of 'critical narrative' can also be used in an ecological sense. The transformative power of a story is made manifest through the dynamic of the narrative itself: through the inter-relations between the story, the story-teller and the listeners. It has the power to change its listeners or to lure them into certain attitudes and behaviours. Attitudes to ecology can thus be developed through memory. Johnson's third category of remembrance, which is 'solidarity in difference', can also be used ecologically. When we remember and identify with others with the kind of solidarity she is advocating, our connection with them becomes so strong that their joys and pain become ours and a spur to transformative action. This is just as true of our solidarity with all of creation as it is with other human beings. A compelling desire for orthopraxis emerges when we remember and identify with both the human and non-human victims of the past.

Whilst Johnson's work on memory is mostly anthropocentric in focus, it can easily be adapted to include solidarity with all of creation and remembrance of those parts of creation which have been victimised or destroyed completely. Moreover, her work also has an eschatological focus. Although she acknowledges the problems with traditional eschatological categories, she nevertheless affirms the importance of hope in a setting which is not solely immanent. Her position is also strengthened by her reluctance to spell out too clearly or precisely how this might occur or what shape our future may take.

Developing Conradie's work on 'material inscription' in the context of Johnson's work on memory, we have a model of our participation in this inscription, when we remember creation and give it back to God. We give back to God the things that are God's, and ask God to bless them. However, this process is not restricted to practices of memory, but can be seen in other liturgies as well. The next chapter will develop this further.

Chapter Eight
Alexander Schmemann

8.1 Introduction

Having shown how Johnson's work lends itself to an eschatological model which is ecologically sound, in this chapter I will examine how the liturgical theology of Alexander Schmemann also provides us with a model of participation in the eschatological transformation of creation. The development of this book has been to examine how *active* participation in God's transformation of the cosmos is achieved by human beings and the rest of creation. Liturgical theology, particularly eucharistic theology, shows how human beings may lift up creation to God, praising God on behalf of all of creation, and are therefore enabled to participate in God's transformation of all things in Christ. So we can participate in the process 'material inscription' and as a result are challenged to act ecologically. The work of Alexander Schmemann will be examined, because his work provides us with helpful hints and suggestions how this participation may occur.

8.2 The Eucharist

Eucharistic theology provides another model for remembrance, which is very similar to the work done by Johnson on the communion of saints. We are forever connected to Christ through the Eucharist. We remember Christ in a way which asserts Christ's presence with us even as we remember.

Denis Edwards suggests that theologians must hold together 'God's capacity to achieve God's purposes and the integrity of the process of evolution. The apparent paradox must be acknowledged.'[1] The paradox he is

1. Denis Edwards, *The God of Evolution: A Trinitarian Theology* (New York/Mahwah, NJ: Paulist Press, 1999), 52.

talking about is that, on the one hand, there is an element of the unpredictable in the processes of evolution and indeed the process of history, which would seem to deny any sense of talking about God's salvation. Yet, on the other hand, there is the significant doctrine of divine transcendence, which suggests that God's divine purposes will not be thwarted.

Edwards is talking here about the paradox which exists when talking about the integrity of creation and the doctrine of the transcendence of God and his salvation. Yet another paradox exists which is intimately related to the one Edwards writes about. This is the paradox of affirming that salvation has already arrived in the person of Christ and yet the consummation of all things has not yet occurred. In Edwards' language we might say that salvation is *occurring* in the processes of evolution, yet the final transformation of all things has not yet *occurred*.

I am suggesting that the Eucharist is a concrete and real expression of this paradox. On the one hand we affirm Christ's presence amongst us in the Eucharist as the risen and ascended one or, to use another title, the Cosmic Christ. On the other hand, we cannot yet talk about creation as having been totally transformed: the eucharistic elements are but the first fruits of the new creation. As Tony Kelly puts it, 'the eucharistic bread and wine provide a *foretaste* of life in a transformed universe'[2] [my italics]. The Eucharist is an embodiment of this great paradox of eschatology: the reality of 'now but not yet'. The transformed creation is still unfolding and the final consummation of all things is still to be fulfilled, yet the crucified, risen and ascended Christ comes amongst us in the Eucharist. This is the paradox of the Christian faith: the processes of history and of the world are real and must be taken seriously and yet we also have great hope that God's salvific action in Christ has already begun and that God's redemption of all things will occur.

The Eucharist is, however, not simply an *expression* of the paradox which Edwards is talking about. There is a sense in which, in celebrating the Eucharist, we contribute to the manifestation of Christ's parousia. In recent sacramental theology there has developed an understanding of Christ as the primordial sacrament and the church as the basic sacrament

2. Tony Kelly, *Touching on the Infinite: Explorations in Christian Hope* (Blackburn, Victoria: Collins Dove, 1991), 135. See also Geoffrey Wainwright, *Eucharist and Eschatology* (Oxford: Oxford University Press, 1981), 104–110. It's important to note, however, that Wainwright is wary of the doctrine of transubstantiation. He is quick to insist that the notion of the bread and the wine themselves being a foretaste of life in a transformed universe must not take away from the understanding of Christ as the *host* of the meal.

of the risen Christ.[3] Rahner describes the church as the 'ongoing presence of Jesus Christ in time and space' and 'as the means of salvation by which God offers his salvation to an individual in a tangible way'.[4] Because of this, the church can be understood as the basic sacrament. It is a sign of God's grace, and, as Rahner suggests, it is an *efficacious* sign of God's grace. Yet the church and the church's sacraments are efficacious 'only to the extent that they encounter man's [*sic*] openness and freedom'.[5] There is nothing magical about the sacraments. They do not operate somehow automatically, but rather through God's grace being received and accepted by the individual.

This active role of the material world finds particular expression in the sacrament of the Eucharist. Not all Christian traditions would agree with the understanding of the Eucharist which I am advocating. I agree with Rahner when he writes that the Eucharist is not a repetition of the event which happened at Calvary, but nevertheless 'what happened then enters into our place and our time, and acquires presence and redemptive power within our own being'.[6] It is our ritualistic celebration of the Eucharist which manifests anew the event which occurred as Christ offered himself as a sacrifice to the eternal God. In our acceptance of the presence of the risen Christ in the sacrament of the Eucharist, in our saying 'Yes' to Christ, then 'all is absorbed into the infinity of God and the love of Christ, everything becomes reconciled, everything opens up, everything finds its solution'.[7] Our response to Christ's presence in the Eucharist means that we participate in the transformation of the world into a place filled with the unifying presence of God.

In Rahner's theology the transformative effect of the Eucharist is not restricted to the individual. The self-communication of God, which finds particular expression in the incarnation, and is manifested in the presence of Christ in the liturgical celebration of the Eucharist, helps fashion the community of believers into an efficacious sign of God's grace. Rahner is clear that the church, as the 'tangible and visible community

---

3. Kenan Osborne, *Sacramental Theology: A General Introduction* (New York/Mahwah, NJ: Paulist Press, 1988), chapters 5 & 6. Osborne identifies Karl Rahner and Edward Schillebeeckx as the two main contributors to this line of thought.
4. Karl Rahner, *Foundations of Christian Faith* (London: Darton, Longman & Todd, 1978), 412.
5. Rahner, *Foundations of Christian Faith*, 414.
6. Rahner, *The Eucharist: The Mystery of Our Christ* (Denville, NJ: Dimension Books, 1970), 13.
7. Rahner, *The Eucharist*, 20.

of believers'[8] does not only point to God's *possible* grace and salvific will. Rather the church, through the celebration of a meal, which manifests Christ's presence, '*is* the tangibility and the permanence of this grace and this salvation'.[9] And if the church is the primordial sacrament of Christ, then through the church the self-communication of God continues to find expression.

8.3 Alexander Schmemann

The concept of the world and its processes, and human beings as a part of all of that, participating in God's transformation of the world finds a particularly clear expression in Eastern Orthodox theology. Liturgy is understood to be at the heart of the Christian experience of life and the vocation of Christians, both individually and ecclesiastically, according to the Orthodox theologian and liturgist Alexander Schmemann.[10] Life needs to be understood as capable of being transformed into the liturgy—the work of Christ. If life can be understood to be sacramental, revealing something of the kingdom of God, then our task as human beings is consciously and deliberately to dedicate that life to God. Life becomes then the work of Christ—the liturgy. If life is seen as a liturgy, which helps bring in the kingdom of God, then we have a model of human participation in the consummation of all things. As I will show, this understanding of life as a liturgy need not be solely anthropocentric, but may also include the participation of all of God's creation.

Schmemann's work is defined most readily as that of a sacramental and liturgical theologian. His theology also clearly reflects his active liturgical life in the Orthodox Church. In that context, I view him not as an ecological theologian *per se*, but rather as offering some hints and suggestions for an ecological eucharistic stance. By examining his work in detail, I hope to show how some of these suggestions help develop a liturgical approach to eschatology. In this liturgical model human beings and all of creation are understood as contributing to the manifestation of the kingdom of God in the world.

---

8. Rahner, *Foundations of Christian Faith*, 426.
9. Rahner, *Foundations of Christian Faith*, 427.
10. Alexander Schmemann, *The Eucharist: Sacrament of the Kingdom* (Crestwood, NY: St. Vladimir's Seminary Press, 1988), 10–12.

8.4 The problem of secularism

To understand the relevance of liturgy to the processes of transforming the world, we need to address some particularly Western issues, Schmemann believes. The gradual acceptance and adoption of Western formulations by the East has resulted in a loss of the richness of the Orthodox position. More than that, however, Schmemann believes that the re-consideration of the traditional Orthodox positions will provide the way forward through many impasses. It's my contention that, in addressing these 'Western' formulations, Schmemann provides us with a model of liturgy which addresses the very question of how we participate in God's action of transforming the world.

A key problem in our contemporary theology, and indeed the contemporary mind, asserts Schmemann, is *secularism*. However, Schmemann has a different perspective on the now widely accepted distinction between the secular and the sacred. Schmemann suggests that secularism is, above all, '*a negation of worship*'.[11] It is the denial, both ontologically and epistemologically, of the true meaning of human beings' existence being found in the worship of God. It is through our worship that on an ontological level God is made manifest to humanity. Worship is the locus of what Schmemann terms the 'true epiphany'[12] of human beings' relation to God. Secularism is the rejection of this definition of human beings as being decisive. Interestingly, and quite helpfully, Schmemann defines secularism as a heresy, not about God, but about *man* [*sic*], because it is the negation of man [*sic*] as 'a worshipping being, as *homo adorans*: the one for whom worship is the essential act which both "posits" his humanity and fulfils it'.[13] Secularism denies the basic definition of man [*sic*] as the priest, who blesses God and who receives the world from God and offers the world to God.[14]

Schmemann acknowledges that this definition of secularism, as the negation of worship, is quite contrary to many definitions. Yet he argues it is a definition quite consistent with the Orthodox tradition and, indeed, the acceptance of this definition provides the way past the impasses that

11. Alexander Schmemann, 'Worship in a Secular Age', in *For the Life of the World: Sacraments and Orthodoxy* (Crestwood, NY: St. Vladimir's Seminary Press, 1973), 118.
12. Schmemann, 'Worship in a Secular Age', 118.
13. Schmemann, 'Worship in a Secular Age', 118.
14. Alexander Schmemann, *The World as Sacrament*, (London: Darton, Longman and Todd, 1966), 16.

have developed within the Western Christian tradition and filtered into Orthodox thinking.

To argue his case for this definition of secularism, Schmemann suggests that he must first show that worship implies a certain understanding not just of the relationship of human beings to God, but also of their relationship to the world. Secondly he must show that secularism rejects this understanding.

Firstly Schmemann argues that the nature and meaning of worship has been defined too narrowly, especially in the Western tradition. He is particularly critical of the tendency in the history of Western theology, to reduce sacraments to the categories of 'form' and 'matter', 'causality' and 'validity'. I will discuss the validity or otherwise of his criticism below, suffice it to say now that Schmemann suggests Western theology excluded the broader liturgical tradition from its theological speculations about the formularies necessary for a 'valid' sacrament.

In this respect Schmemann is critical of Western Christianity because of its focus on the 'consecratory formula' and the consequent separation of the Eucharist from the rest of the liturgy and thus from the church, from its ecclesiological essence and meaning. He suggests that the church then loses its understanding of itself not only as the 'dispenser' of the sacraments, but also as their very object: 'they represent her fulfilment of herself in "this world" as the sacrament of the kingdom of God, which "has come in power"'.[15] Western Christianity moved towards an understanding of symbol as something not only distinct from reality, but in fact contrary to it. Indeed Schmemann maintains that the emphasis of Roman Catholicism on the 'real presence' of Christ in the eucharistic gifts grew out of a concern that this presence would only be understood as symbolic. This concerned the Roman Catholic Church, he maintains, because of the distinction that was made between symbolism and reality.[16]

15. Schmemann, *The World as Sacrament*, 29.

16. Schmemann, *The World as Sacrament*, 30. Schmemann's criticism of the Western Church is not entirely fair, in the sense that it is too broad a criticism. Walter Kasper comments that Schmemann's criticism is really only valid 'as far as the traditional theology of the handbooks is concerned, but does not begin to do justice to the great scholastic theology'. Walter Kasper, *Sacrament of Unity: The Eucharist and the Church* (New York: Crossroad, 2004), 170. But Kasper does suggest that we need to recover an emphasis on the universal, cosmic dimension of the Eucharist. He writes that when the bread and the wine, as the gifts of creation and the fruit of human work 'are brought into the eucharistic event, the eschatological transformation of all reality is in a certain sense accomplished in them even now'. Kasper, 170. Kasper reminds us (and in doing

What Western Christianity forgot, or overlooked, Schmemann suggests, is the liturgical tradition which understood the character of the world as *sacramental*. This understanding of the world is expressed in worship; indeed worship derives its very *raison d'être* from this belief. Schmemann defines this sacramental character of the world as meaning:

> that the world, be it in its totality as cosmos, or in its life and becoming as time and history, is an *epiphany* of God, a means of His revelation, presence and power.[17]

---

so disarms some of Schmemann's criticism of the Western Church) that Teilhard de Chardin had discovered anew the cosmic dimension and 'irradiation' of the Eucharist. The Eucharist indicates the direction of the cosmic movement, which is towards the divinisation of the world, which it also anticipates. More recently Denis Edwards has built on the work of Teilhard and developed his eucharistic theology in this direction, suggesting that every 'Eucharist is both sign and agent of the transforming work of the risen Christ in the whole of creation'. Denis Edwards, 'Celebrating Eucharist in a Time of Global Climate Change', in *Pacifica* 19 (February 2006), 13.

17. Schmemann, 'Worship in a Secular Age', 120. Some of the work of the Anglican Roman Catholic International Commission does exactly this, as too does the eucharistic theology of Jean-Marie Roger Tillard, who was heavily involved in the writing of the Anglican-Roman Catholic International Commission (ARCIC) Statement on the Eucharist. The 1971 ARCIC Statement on 'Eucharistic Doctrine' certainly bears out Kasper's comments, that Western sacramental theology is more developed than Schmemann allows for. Whilst acknowledging that Christ's redeeming sacrifice on the cross and resurrection 'took place once and for all in history', the Commission nevertheless stated that the Eucharist has been given by God to his church 'as a means through which the atoning work of Christ on the cross is proclaimed and made effective in the life of the Church'. Anglican-Roman Catholic International Commission, *The Final Report: Windsor, September 1981* (London: SPCK and Catholic Truth Society—CTS, 1982), 13–14. The Commission based this statement on an understanding of the Eucharist as a memorial (*anamnesis*) of Christ's sacrifice, by which they meant 'the making effective in the present an event in the past'. ARCIC, 14. In the 1979 'Elucidation' of the Statement the Commission stated that the word *anamnesis* expressed well the reality of 'the once-for-all event of salvation [becoming] effective in the present through the action of the Holy Spirit'. ARCIC, 19. The ARCIC documents help answer Schmemann's criticism that Western theology had made a distinction between symbolism and reality. The Commission was convinced of the need to emphasise sacramental realism. The Commission wrote that the 'Lord who thus comes to his people in the power of the Holy Spirit is the Lord of glory'. ARCIC, 16. The Eucharist is therefore much, much more than simply a reminder of Christ's sacrifice. Through the transforming power of God's Spirit, the bread and the wine '*become* the heavenly manna and the new wine, the eschatological banquet for the new man [*sic*]' [my italics]. ARCIC, 16. Moreover, as Tillard reminds us, no celebration of the Eucharist occurs in isolation. Every Eucharist is celebrated not only in the presence of a diversity of people, with a variety of social status and sometimes race and language, but also 'in communion with the

The world not only suggests that God could be the cause of its existence, but also that the world itself is 'an essential means both of knowledge of God and communion with Him'.[18] The world's ultimate destiny, and its true nature, is true communion with God, though in a transformed state.

In this context, worship is the means by which human beings experience what we intuitively long for: communion with God and true knowledge of God and therefore knowledge of the world. Worship gives us 'communion with God and therefore communion with all that exists'.[19] What Schmemann terms the 'natural sacramentality' of the world finds expression in worship. It is through communion with God that human beings, who, it needs to be remembered, are not somehow separate from the world, find ultimate fulfilment.

From a *Religionswissenschaft* (history of religion) perspective, worship is a primordial and universal phenomenon. However, the Christian theologian need not understand this negatively. Indeed the reality of the continuity of Christian worship with worship in general need not deny the uniqueness of Christianity. Its uniqueness lies not at all in discontinuity with previous forms of worship, but rather in the person of Christ. Indeed, suggests Schmemann, Christ fulfils that continuity, by being 'the ultimate "epiphany" of man [*sic*] as worshipping being'.[20] Christ is the fulfilment of God's manifestation by the means of the world. He is the completion of the world's sacramentality.

Any newness in Christian worship is therefore found in the person of Christ. Schmemann is careful to point out that any discontinuity that does exist in Christian worship is not a negation of previous forms of worship, but rather comes as a result of the rejection of Christ by the world. In rejecting Christ the world rejected its own fulfilment and destiny. If

---

saints in heaven, even with all those who will gather at the Lord's table until the parousia'. Jean-Marie Roger Tillard, *Flesh of the Church, Flesh of Christ: At the Source of the Ecclesiology of Communion* (Collegeville, Minnesota: The Liturgical Press, 2001), 108. In the context of this communal celebration, 'each and all together flow into the source of the sacrifice of holy life, the paschal sacrifice of Christ made present'. Tillard, 109. The Eucharist then becomes a celebration of our solidarity with one another, whilst we also enjoy the moment of greatest intimacy with the Lord, since, as Tillard puts it, 'one becomes his body'. Tillard, 28. From an ecological perspective, and using the Orthodox theology of Schmemann, I am adding to both Tillard's and Schmemann's work by suggesting that all of creation is caught up in each eucharistic celebration, so that the Eucharist also becomes an expression of our solidarity with all of God's creation.

18. Schmemann, 'Worship in a Secular Age', 120.
19. Schmemann, 'Worship in a Secular Age', 120.
20. Schmemann, 'Worship in a Secular Age', 122.

the Incarnation is the basis of all Christian worship, then the Cross and Resurrection are what Schmemann calls its 'true content'.[21] Following the pattern of Christ, the things of this world must be lifted up, taken out of this world, in order to ascend with Christ to his table in the kingdom. In the context of the Eucharist, which I will discuss in more detail below, the symbols of this world, the bread and the wine, are taken out of this world and transformed into the Body and Blood of Christ.

Schmemann's use of the concept of taking the things of the world out of the world is unfortunate, in the sense that it potentially evokes in the reader thoughts of neo-Platonism. If the things of this world are taken out of the world, is the world itself actually valued? It's important to note, though, that for Schmemann this is not a negation of the things of this world, or of the world as a whole, any more than Christian worship is a negation of previous forms of worship. Rather all things are *made new* in Christ. The key to understanding this is *remembrance*. Because Christian worship is remembrance of Christ 'in the flesh',[22] then it is also the remembrance (in the sense of anticipation and expectation) of his kingdom. The Christian liturgy is always cosmic and historical; it assumes all creation into Christ and it assumes all time into Christ. Both all of creation, and all of time, are then made new and so worship is fundamentally eschatological. Worship is therefore not an act of piety (which is how it is so often viewed, particularly by the 'secular' world) but rather an all-embracing world view. It is an act with 'cosmic, historical and eschatological dimensions'.[23] Significantly, the cosmos, including the 'historical dimensions' of the cosmos, are all part of the transformation.

To develop Schmemann's ideas further in an ecological direction, this understanding of all of creation being lifted or offered up to God means that all of creation is caught up in each eucharistic celebration. The Eucharist itself becomes an expression of our solidarity with all of God's creation. This in turn is a powerful reason to act in an environmentally responsible way. Moreover, worship is an active form of participation by human beings in God's eschatological transformation of the world. By offering worship to God, and by including all of creation in what is offered to God, we participate in the transformation, because the things of the world are raised up to God. I will discuss this concept further as I examine Schmemann's theology in more detail.

21. Schmemann, 'Worship in a Secular Age', 122.
22. Schmemann, 'Worship in a Secular Age', 123.
23. Schmemann, 'Worship in a Secular Age', 123.

Secondly, Schmemann suggests that secularism rejects this understanding of the world, and of worship. Secularism, he is quick to point out, is actually not the same as atheism. Atheism is essentially the rejection of the possibility of the existence of God. Secularism, on the other hand, may well accept the idea of God. However, what the secularist 'emphatically negates is precisely the sacramentality of man [*sic*] and world'.[24] The secularist fundamentally rejects the possibility of epiphany. Schmemann reiterates his understanding, that an acceptance of epiphany is an awareness of the primordial intuition that the world and everything in it not only have their cause of existence elsewhere, but also are themselves the 'manifestation and presence of that *elsewhere*'.[25] Moreover, it is from that elsewhere, which Christians understand to be God the Holy Trinity, that the life of the world and everything in it is ultimately fulfilled. If we are cut off from the source of that life, we are left only with 'darkness, absurdity, and death'.[26]

Schmemann's work reveals an almost incredible irony in contemporary society: a secularist may in fact be quite religious! The critical issue is not belief in God, *per se*, but rather the rejection of the possibility of the world making manifest the presence of God. The idea that the divine might be present in the world and the things of the world is rejected. This is despite the fact that the secularist may indeed accept the existence of God, the possibility of creation having come from God, and the possibility of God's intervention in the world. He or she may even without difficulty, says Schmemann, believe in a concept of the immortality of the soul, as well as develop ecclesiastical programmes with a focus on theology. Crucially, though, the secularist does not allow for the possibility of epiphany. Symbols, when used, are seen only as 'audio-visual aids, for communicating ideas'.[27] According to the secular world-view, even if that world-view is espoused by a religious person, symbols are viewed merely as illustrations of concepts and ideas.[28]

In a profound insight Schmemann views secularism not as something which has arisen in isolation from Christianity, but which is rather Chris-

24. Schmemann, 'Worship in a Secular Age', 124.
25. Schmemann, 'Worship in a Secular Age', 124.
26. Schmemann, 'Worship in a Secular Age', 124.
27. Schmemann, 'Worship in a Secular Age', 125.
28. Puritans may well fit into this category of religious secularists. At its logical extreme Puritanism is a form of dualism, 'a separation of the cerebral and the spiritual from the bodily and the material. In its intense rationalism it is also highly anti-intuitive.' Porter, *The New Puritans*, 21.

tianity's 'step-child'. Because of this, secularism is best described as a heresy, which Schmemann defines as a distortion of the truth, an exaggeration, the affirmation of one element of the faith at the expense of the others, the rejection of the totality or catholicity of the truth. He sees the seeds of contemporary secularism, and therefore many of the problems confronting Christianity, lying in our Christian history. Unlike the heresies of the patristic age, which he suggests arose out of Christianity's encounter with Hellenism, secularism is the result of the deep metamorphosis of Christianity itself.

The example Schmemann chooses to use to illustrate his point is that of the debate about the presence of Christ in the Eucharist. At the end of the twelfth century (I note that this is long before the debates of the Reformation about the 'Real Presence' of Christ in the Eucharist) a theologian named Berengarius of Tours was condemned for teaching that the presence of Christ in the Eucharist was 'symbolic' rather than 'real'. Tragically, from Schmemann's point of view, a view which I share on this matter, the Council which condemned him simply reversed his formula and said that since Christ's presence in the Eucharist was real, it was not mystical. What both sides accepted was that 'reality' and 'symbolism' were mutually exclusive.[29] I would suggest that perhaps the Council failed to realise the consequences of this acceptance, because this was in fact the collapse of a fundamental Christian mystery: the holding together of truth and symbolism, 'of the reality of the symbol, and of the symbolism of reality'.[30] Schmemann's example illustrates the loss of the comprehension of creation in terms of its 'ontological sacramentality'.[31] From this position of the loss of the true understanding of Christian symbolism, the false dichotomy developed between 'natural' and 'supernatural'. The result of this dichotomy is that 'the world ceases to be the "natural" sacrament of God, and the super-natural sacrament to have any "continuity" with the world'.[32]

The terrible result of this dichotomy is, in the end, the emergence of secularism. But secularism is the result of a Christian heresy, because the separation of truth and symbolism has resulted in an understanding of the world being 'grace-proof'. Whether one accepts the world in this state, as Schmemann suggests Western enthusiasts of 'secular Christianity' do, or rejects it utterly, as in the case of what he terms 'the "Super-Orthodox"

29. Schmemann, 'Worship in a Secular Age', 129.
30. Schmemann, 'Worship in a Secular Age', 129.
31. Schmemann, 'Worship in a Secular Age', 129.
32. Schmemann, 'Worship in a Secular Age', 129.

prophet of apocalyptic doom'—though the criticism could apply to all apocalypticists—the premise is the same. Both positions accept the false dichotomy.

8.5 Redemption of the 'fallen' world

Schmemann has an interesting perspective on the 'fallen'[33] nature of the world and Christ's redemption of it. The problems arose when creation turned away from God, and rather than looking to God for life and meaning, looked solely to itself for meaning. Using Schmemann's concept gives us a helpful interpretation of the story of the Garden of Eden. The symbolism of the forbidden fruit is that Adam and Eve ate that which was not blessed by God, not given by God. 'It is an image of the world loved for itself, and eating it is the image of life understood as an end in itself.'[34] The result is that human beings see the world as opaque, rather than revealing God's presence. The sin of human beings, Schmemann suggests, has been precisely to make the world an end in itself, rather than a symbol of God's kingdom. Further to that discussion, in a very perceptive and interesting reflection, Schmemann suggests that the 'original' sin was not primarily that human beings disobeyed God, but rather that they ceased to see their 'whole life depending on the whole world as a sacrament of communion with God.'[35] Human beings distorted the balance between the material and the spiritual, making the world material, whereas they were supposed to 'have transformed it into "life in God", filled with meaning and spirit.'[36] Thus human beings did not prefer the world to God, but rather preferred the *material* to the *spiritual*.

33. In examining Schmemann's theology of the 'fall' and his use of the word 'restore' I note the problems raised by the issue of temporality. For a helpful discussion of the question of temporality and the role of human beings in the 'fall' see Nicola Creegan, 'Being an Animal and being made in the Image of God', in *Colloquium: The Australian and New Zealand Theological Review* 39.2 (November 2007): 185-203. Creegan argues that, in the light of the theory of Evolution and DNA showing an inseparable connection between human beings and animals, we need to re-think the time-line of the 'fall' and allow for a concept of a progressive 'fall'. Moreover, she argues that evolutionary theory suggests that, 'Nature itself, although perfect in some way, is itself already fallen or at least incomplete before humanity arrives.' Creegan, 201.
34. Schmemann, *The World as Sacrament*, 17.
35. Alexander Schmemann, *For the Life of the World: Sacraments and Orthodoxy* (Crestwood, NY: St. Vladimir's Seminary Press, 1973), 18.
36. Schmemann, *For the Life of the World*, 18.

In Christ, however, the world has been 'restored'[37] to God, and the church makes this reality manifest. In this respect Schmemann's eucharistic theology reminds those of us from a Western tradition of the richness of the Eastern Orthodox understanding of symbolism in the Eucharist, and it is particularly helpful in an eschatological context. Schmemann suggests that the purpose and function of a symbol is 'to *manifest* and to *communicate* what is manifested. We might say that the symbol does not so much "resemble" the reality that it symbolises as it participates in it, and therefore it is capable of communicating it in reality.'[38] The Eucharist, therefore, as a sacrament of the kingdom is not merely a memorial meal (understanding 'memorial' in a limited way in this context) but rather actually communicates the reality of the kingdom and *participates* in its manifestation. The church's task, in the celebration of the Eucharist, which is the sacrament of the kingdom, is to 'fulfil this symbol, to realise it as the "most real of realities". We can therefore say that the symbol reveals the world, mankind [*sic*] and all creation as the "matter" of a single, all-embracing sacrament.'[39] The reality of the kingdom is made manifest in the sacrament. Significantly, and most importantly for Schmemann, the symbol is not merely a simple sign of the deeper reality of the kingdom, in the sense of 'standing for' the reality or 'illustrating' it. Rather, the symbol unites the empirical (the visible) and the spiritual (the invisible) *epiphanically*.[40]

Through this understanding of the world as a sacrament, the false dichotomy between 'reality' and 'symbolism' is overcome. Reality is experienced, above all, as the *fulfilment* of the symbol. Indeed the church's task in this world is to fulfil the symbol. Undergirding Schmemann's theology is an appreciation of the early church's belief that the kingdom was already at hand, and that it had appeared, but remained hidden and unseen by this world. The strength of the early church's liturgy was its eschatological nature: the future parousia was experienced in the present. The liturgy, particularly the liturgy of the Eucharist, was understood 'as the epiphany of that which is to come, as communion with the "world to come"'.[41] Thus the symbol unites heaven and earth, what Schmemann calls the two dimensions of the church.

37. See my comments on the question of temporality above, at footnote 96.
38. Schmemann, *The Eucharist*, 39.
39. Schmemann, *The Eucharist*, 40.
40. Schmemann, *The Eucharist*, 39.
41. Schmemann, *The Eucharist*, 43.

The manifestation of the kingdom of God is participatory, in the sense that the priest and indeed the whole people of God participate in 'the blessing of the kingdom of God'.[42] Schmemann understands this to mean that 'we *proclaim* it to be the goal of the sacrament—of pilgrimage, ascension, entrance—that now begins'[43] [my italics]. At the end of the Orthodox eucharistic prayer the people respond with 'Amen'. Schmemann suggests that the more usual translation of 'so be it' does not portray the true strength of the word. What is really meant is 'Yes, this is so, and *let* it be so'.[44] Schmemann goes on to say that this fuller translation implies action from the people too. 'With this word the ecclesial assembly concludes and, as it were, *seals* each prayer uttered by the celebrant, thereby expressing its own organic, responsible and conscious participation in each and every sacred action of the Church.'[45] In this time of global ecological crisis this participatory action needs to be understood as including but also going beyond liturgical action. The connection between the liturgical life of the church and its members, and the members' practical, everyday life, needs to be reaffirmed. The sacramental understanding of the world, which Schmemann believes in and promotes, lends itself to reaffirming this connection. If the false dichotomy made between the sacred and the secular is finally rejected, then it is obvious that the Christian's action in the world ought to reflect his or her Christian convictions. Clearly this extends to ecological action.

Thus in *proclaiming* the kingdom of God, we play our role in bringing it about. This is a more active role than the passive one implied by Conradie's understanding of 'material inscription'. Schmemann's liturgical theology gives us a framework in which the role of human beings and, as I will develop, all of creation, is active and participatory. The kingdom of God does not come about whilst human beings and all of creation stand by watching idly. To use an image from Jesus' parable of the labourers in the vineyard, we all are called to work in the vineyard.[46] Even if, at the last hour, we finally do something about climate change, it is not in vain.

42. Schmemann, *The Eucharist*, 47.
43. Schmemann, *The Eucharist*, 47.
44. Schmemann, *The Eucharist*, 48.
45. Schmemann, *The Eucharist*, 48.
46. Matthew 20:1–16.

8.6 A theology of blessing

The Orthodox understanding of symbolism, to which Schmemann seeks to return, has the potential to provide us with the subjective side of a model of the eschatological transformation of all things. If we put aside the false dichotomy between the sacred and the profane, and return to an understanding of the world as sacrament, then a theology of blessing may provide a way forward. I will briefly look at Schmemann's work on 'holy water' and Baptism and Confirmation before discussing in more detail his theology of the Eucharist, as examples of how a theology of blessing may help us.

Schmemann uses the example of blessing water in order to illustrate his point about the development of a false dichotomy. It may mean essentially one of two things. The first understanding, which even the atheist would recognise, if not accept, is the transformation of something profane into something sacred. The water becomes holy water, as opposed to ordinary water. In this understanding the act of blessing says nothing about water, or the world and the things of the world. Rather the new function of the holy water completely overshadows the old. Indeed the holy water is completely opposed to the 'mere' water. Any meaningful continuity of the sacred water with the mere water is lost. This understanding of blessing arises from the secularism which Schmemann so vehemently opposes.

The second understanding is quite different. The same act of blessing, viewed from what Schmemann thinks of as an authentic Christian position, before the emergence of the heresy of secularism, means the 'revelation of the true "nature" and "destiny" of water, and thus the world—it may be the epiphany and the fulfilment of their "sacramentality"'.[47] Blessing restores the water to its proper function and so, in being revealed as full, adequate water, becomes a means of 'communication and knowledge of God'.[48]

This restoration of water to its true function is a helpful perspective. All of creation longs to be restored to that purpose, for which it was created. That is, it longs to be restored to its role of praising God, and in doing so, revealing God, or *unveiling* God, to use the language of the Apocalypse. This is the true role of all of God's creation. When all of creation is engaged in its true role of revealing God, it is participating in the transformation of all things. All parts of creation, then, are valued in an ecological and

47. Schmemann, 'Worship in a Secular Age', 132.
48. Schmemann, 'Worship in a Secular Age', 132.

eschatological sense, because they play their part in revealing the divine. This is a significant reason to avoid wantonly ruining parts of creation. We may well be preventing, or, at the very least, hampering creation in its role of revealing God. A real sense of orthopraxis may emerge from the simple realisation that creation reveals God's presence.

Unfortunately, that understanding of the blessing of water has been almost entirely excluded by the almost magical comprehension of the supernatural. Yet, contrary to that supernatural understanding, true worship, or blessing, has as its very function, the power to transform and change. Casting aside the asphyxiating heresy of secularism, true worship transfigures the world. There exists continuity between religion and life, through the deep communication with the Divine that occurs in an act of true worship. Schmemann is convinced that the world is desperate in its need for Sacrament and Epiphany. Extending Schmemann's thoughts a little further here, he gives us a model of the new creation, in which the old creation is not negated, or annihilated, but rather transformed. Based on remembrance, the new creation is the total fulfilment of the old, transfigured in worship.

The blessing of water is a symbol therefore of what happens to the whole world. Indeed, to move on to his work on Baptism, Schmemann points out that the baptismal water 'represents the matter of the cosmos, the world as life of man [*sic*]. And its blessing at the beginning of the baptismal rite acquires thus a truly cosmic and redemptive significance.'[49] To bless, as Schmemann reminds us, means to give thanks. Through this act of thanksgiving, the blessing of the water in Baptism, we acknowledge the true nature of creation, given to us by God. In offering the water up to God, we return it to its initial and essential meaning. 'Once more the world is proclaimed to be what Christ revealed and made it to be.'[50] But, crucially, it is this act of blessing, of giving thanks, which restores the things of this world and the world itself to their true meaning, which is 'the gift of God to man [*sic*], and the meaning of man's [*sic*] communion with God.'[51]

All of this happens in Christ, so that the liturgy is the work of Christ. We put aside the desire to make life an end in itself, and we unconditionally dedicate that life to Christ. Through our work (the liturgy), which becomes Christ's work, the world then truly becomes a sacrament of the

49. Schmemann, *The World as Sacrament*, 88.
50. Schmemann, *The World as Sacrament*, 89.
51. Schmemann, *The World as Sacrament*, 89.

presence of Christ, 'the growth of the kingdom and of life eternal'.[52] The world is not transformed into some mysterious 'other' world, as though our experience in this one were second rate or insignificant. It is precisely this world, which, through our work of thanksgiving, becomes what it is truly meant to be, a sacrament of Christ's presence. This understanding of the world potentially has profound practical consequences. If the world is indeed a sacrament of Christ's presence, then it can never again be viewed as expendable. It cannot be understood as being able to be abused for greed and then cast aside. Understanding our liturgical life as the work of Christ, which helps transform the world, influences our action in the world.

8.7 Confirmation

In the Orthodox tradition, a tradition shared in this respect by some other denominations, such as the Anglican, Lutheran and Roman Catholic churches, the sacrament of Confirmation is linked with the sacrament of Baptism. Schmemann suggests that Confirmation is an integral part of the baptismal liturgy and needs to be understood as the fulfilment of that liturgy. The invocation of the Holy Spirit marks the manifestation of the 'last things', so that the kingdom of God is present in this world, through the sacrament of the church. Schmemann's theology is distinctly catholic[53] and ecclesial at this point. The gift of the Holy Spirit to the individual, and the dedication of that individual's entire life to God, is not separated from the transformation of the church itself. In the anointing of the individual with the Holy Spirit, he or she is dedicated to a new life. 'The whole man [*sic*] is now made the temple of God, and his whole life is from now on a *liturgy*.'[54] The whole Confirmation liturgy is a symbol of what the whole of life, and the whole of God's church, is about. There is no separation, in an ontological sense, between the Confirmation of an individual, and the dedication of his or her entire life to God, and the transformation of the church, and indeed the world, into a liturgical act. What happens in the Sacrament of Confirmation is what we desire for the world itself, as Schmemann puts it:

52. Schmemann, *The World as Sacrament*, 91.
53. I'm using the word 'catholic' here in its broader sense of 'universal'.
54. Schmemann, *The World as Sacrament*, 92.

> Each ounce of matter belongs to God and is to find in God its fulfilment. Each instant of time is God's time and is to fulfil itself as God's eternity. Nothing is 'neutral'. For the Holy Spirit, as a ray of light, as a smile of joy, has 'touched' all things, all time—revealed all of them as precious stones of a precious temple.[55]

What happens in the sacrament of Confirmation is a symbol of what happens in the sacrament of the church, and the sacrament of life: the Holy Spirit, the *ruah* of God, enters our lives and 'makes us available for divine action, filling everything with joy and hope'.[56] All of life becomes a liturgy—the work of Christ. Through our sacramental actions we return the matter of life to God, thus, through Christ, transforming it into the material of the kingdom of God.

This understanding of all of life becoming a liturgy is not restricted to the formal sacraments of the church nor to the activities of human beings. Rather all of life, and all of life's processes, can be understood as sacramental actions, which make up a *cosmic liturgy*. This cosmic liturgy of all of creation restores the world to what it was once was, and is meant to be, a sacrament of God's presence. I will further discuss later in the chapter how this might be so. It's important to remember here that, if all of life makes up the cosmic liturgy, then the destruction of the environment clashes loudly with the otherwise beautiful music of the liturgy. It is as if, in the middle of a beautiful classical symphony, a jack-hammer starts up in the concert hall.

8.8 Schmemann's eucharistic theology

Schmemann's eucharistic theology depends on the ontological understanding of the world as being sacramental, which I discussed above. In the Orthodox tradition the Eucharist occupies a special place and is understood as the 'sacrament of the kingdom'.[57] Schmemann provides us with an understanding of the Eucharist, which succinctly expresses the participatory role human beings and indeed all of creation have to play in God's consummation of all things. However, it's important to remember that Schmemann's eucharistic theology, like his theology of Baptism

55. Schmemann, *The World as Sacrament*, 93.
56. Schmemann, *The World as Sacrament*, 93.
57. A Schmemann, *The Eucharist*, 27.

and Confirmation, is not limited to the church's rites. Schmemann is concerned to avoid the trap of thinking of the church's role as being sect-like, in the sense of isolated from the world. The constant temptation for Christians, as noted above, is to accept the distinction between the sacred and the secular. The temptation is powerful, because it's ostensibly 'easier to live and breathe within neat distinctions between the sacred and the profane'.[58] But Schmemann avoids such distinctions and sees the world as the material of the Eucharist, an all-embracing Eucharist. Any celebration of the Eucharist, in this world-view, is a part of the liturgy of the world, in which human beings convert creaturely life into participation in divine life.[59] Thus all sacraments, including the Eucharist, are cosmic; they pertain to the restoration of the entire cosmos, all of creation, to God.

Schmemann's work on the Eucharist examines each part of the Orthodox celebration of the Eucharist and then relates it to the underlying belief that the Eucharist is the sacrament of the kingdom. For the purposes of this work, I shall restrict my discussion to those points directly connected to my proposal, namely, that we need a model of active participation in God's consummation of all things, which includes the participation of all of creation, not simply human beings.

8.8.1 The sacrament of entrance

The opening part of the Orthodox liturgy, the Entrance, is significant because the world is being brought to God. The church enters into the kingdom of God, and brings with it the world, for the sake of its salvation. In that sense, Schmemann points out, the entire celebration of the Eucharist is an entrance into God's kingdom. If there is a separation from the outside world at this point, it is not so that the church becomes a sect, but rather that the church deliberately enters into God's presence and offers up the world; the world is gathered up in the church's liturgy and entrance.

The key movement at this point of the liturgy is ascent. We ascend to the same place Christ ascended: the heavenly realm. It is here that the world is offered to God. Every celebration of the Eucharist, therefore, is a celebration in heaven. Because we ascend to where Christ ascended, we can speak of the heavenly nature of each eucharistic celebration. The people have entered the heavenly realms and stand before the holy, so that they might offer up the world.

---

58. Schmemann, 'Worship in a Secular Age', 130.
59. Schmemann, *The Eucharist*, 34.

The criticism might be made of Schmemann at this point that his theology is in fact secularist, that is, that it falls into precisely the trap of separating the sacred from the profane. This is because the altar, on which the eucharistic sacrifice is offered, might be seen as sacred, in the sense of having been sacralised, and marked as different to the rest of creation. Importantly Schmemann is quick to remind his readers that any consecration in the church is not the consecration of sacred objects, as opposed to the profane, but rather their restoration to their symbolic nature, their sacramentality. Thus the altar, just like the whole world, is restored to its original symbolic meaning, and to 'the ultimate aim of being',[60] because the church has ascended into the heavenly realms and stands before the holy, offering the whole world. Our veneration is never related to 'matter', as Schmemann puts it, but to the presence of God, which is revealed or manifested in the symbol.

8.8.2 The sacrament of the faithful

Schmemann helpfully reminds us that each celebration of the Eucharist, and indeed the entire life of the church, conjoins what at first appear to be two contradictory affirmations. On the one hand the church is directed to the entire world, to the whole of creation. Yet, on the other, the church also affirms the saving love of Christ for each human being, each individual. The church, and the Eucharist in particular, is 'simultaneously directed to the whole—to the entire creation, the whole world, all mankind—and to each unique and unrepeatable human person'.[61]

The result of this is that the fulfilment of the world lies in it becoming the means to fullness of life for each person, and I would add each creature. It also means that each person was created for the world, 'to surrender himself [*sic*] for "the life of the world"'.[62] This is a necessary corrective for the Western tradition with its heavy emphasis on 'individual salvation', as if that salvation were somehow separate from the world.

If these two apparently contradictory foci of the church and the Eucharist can be held together, then the liturgy of the church can be simultaneously the cosmic sacrament of salvation for the world, as well as the prayers and offerings of the people as individuals. Individual prayers are then not isolated from the greater cosmic celebration—indeed they contribute to it. When we offer up 'individual prayers' for our loved ones,

60. Schmemann, *The Eucharist*, 61.
61. Schmemann, *The Eucharist*, 82.
62. Schmemann, *The Eucharist*, 83.

praying for example for healing and peace, then these prayers are a part of the greater cosmic liturgy, offered up for the world as a whole. The prayer for the healing of individuals is not isolated from the prayer for the healing of the world. All prayers have at their core, as the church's business, 'the incessant acquisition of the Holy Spirit and growth into the fullness of Christ, who lives in her'.[63] And God gives the Spirit without measure, responding to each prayer, each act of liturgy, as part of the great cosmic liturgy, bringing all of creation to fullness and perfection, to abundant life.

Another way of expressing the corporate and individual nature of the church's worship is found in the structure of the church itself. Using the Orthodox tradition as an example, Schmemann argues that each part of the church is able to live as a part of the whole, and yet also wholly. Thus no parish by itself is 'the whole' or the 'church', as tends to be asserted by Protestant congregationalism. The parish is only a part of the church, and, in the Orthodox tradition and others, is linked through the Bishop to the fullness of the church. Yet neither is a parish subsumed into the whole, thereby losing its individual identity and expression. Rather, especially through the gift of the Eucharist, which is a gift of the whole church to the individual parish, the 'parish partakes of the "whole Christ", receives all the fullness of the beneficent gifts and identifies itself with the Church'.[64]

This discussion of ecclesiology leads Schmemann to a timely reminder of one of the fundamental principles of eucharistic worship, that each individual celebration of the Eucharist is, at the same time, 'the offering, presence and coming of the *whole Christ*',[65] in whom all parts are overcome by the whole. Each time the Eucharist is celebrated, it is part of the whole work of Christ, of building up the kingdom of God.

This juxtaposition of apparently contradictory concepts has ramifications for an ecological interpretation of the sacrament of the Eucharist. Each celebration of the Eucharist is doing the work of Christ. Each celebration of the Eucharist is a part of a greater liturgy, the liturgy of the whole church. But, to push Schmemann's thoughts a little further, each celebration of the Eucharist is a part of the greater cosmic liturgy, in which the whole of creation plays its part in blessing God. I will discuss this concept in more detail later, but suffice to say for now, each act of liturgy, each celebration of the Eucharist, allows human beings to participate in God's transformation of the world. Praising God, and including all of creation in

63. Schmemann, *The Eucharist*, 90.
64. Schmemann, *The Eucharist*, 98.
65. Schmemann, *The Eucharist*, 99.

the praise, is not an optional extra in life. Our praise of God, our blessing of God, and our bringing of all of creation to God, is one *part* of a greater cosmic liturgy, occurring all the time.

8.8.3 The sacrament of offering

Schmemann's work on the sacrament of Offering begins with a helpful insight into the whole theology of sacrifice. He suggests that, whatever we think of the Jewish system of sacrifices, and indeed the plethora of pagan sacrificial rituals, they reveal human beings' 'primordial, indestructible thirst for God'.[66] From a Christian perspective, however, all these sacrifices were unable to deal with sin once and for all. This only occurs with the sacrifice of Christ on the cross, who 'offered himself as a sacrifice for the life of the world'.[67] Using the language of the letter to the Hebrews (Heb. 9:28) Schmemann describes this sacrifice as single, all-encompassing and unrepeatable.

Within the context of this theological understanding of Christ's sacrifice on the cross, Schmemann then discusses how our lives, and the life of the whole world, are offered up to God, through Jesus Christ. Crucially this offering is not a new sacrifice, but an offering of Christ himself. This is made clear in the offering of the bread and wine. The bread and the wine are identified as symbols of the sacrifice of Christ, but, remembering Schmemann's theology of symbols, the term symbol must not be understood in opposition to 'reality'. In offering the bread and the wine we are offering Christ, but the sacrifice of Christ 'encompasses all our sacrifices, our offering of our very selves to God'.[68] Thus all of us, individually and together (not forgetting the rest of creation either) are offered to God, through Christ. In this offering, or *commemoration* as Schmemann also calls it, the partition between the living and the dead is overcome. Both the dead and the living are offered to God. We give God's own creation back to God, in order to make all of creation a participant in the life of the kingdom of God.

Everything is referred to God's memory. The prayer that God would remember everything is at the heart of all of the church's worship. Schmemann reminds us that the Old Testament understanding of memory refers to God's attentiveness to his creation, that is, God's providential love. Life

66. Schmemann, *The Eucharist*, 102–103.
67. Schmemann, *The Eucharist*, 104.
68. Schmemann, *The Eucharist*, 110.

is then understood as 'abiding in the memory of God, and death the falling out of this memory'.[69] God's memory is real and abiding.

Our memory of God is the particular gift and task of human beings. Schmemann raises a fascinating issue here. He asserts that,

> if everything in the world witnesses to God, declares his glory and renders him praise, then only man [*sic*] "remembers" him and, through this memory, through this living knowledge of God, comprehends the world as God's world, receives it from God and raises it up to God. To God's remembrance of him, man [*sic*] answers with his remembrance of God. If God's remembrance of man [*sic*] is the gift of life, then man's [*sic*] remembrance of God is the reception of this lifecreating gift, the constant *acquisition* of and increase in life.[70]

Do other creatures remember God? Perhaps it is possible that in some limited way (limited from our human perspective) they do. Yet Schmemann asserts that memory is the particular gift and responsibility of human beings. The origin of sin is that human beings forgot about God, and so God ceased to exist for them. This forgetting of God has destructive consequences for human beings, and, I would add, for God's creation. Memory ceases to be the life-creating power, which it is when understood as remembrance, and becomes instead mere 'knowledge of death and the constant tasting of mortality'.[71]

The restoration of memory, on the other hand, restores it as a life-creating power. God's remembrance of human beings is fulfilled in our remembrance of God. This is a specific task for human beings. We are to remember Christ. This memory of Christ is not mere knowledge of the one remembered, but the acceptance of his living presence in our lives. Thus, whilst salvation is accomplished in Christ, we remember Christ, who is alive, and 'this remembrance becomes our entry into *his* victory of time, over its collapse into "past", "present" and "future"'.[72] It is an entry into God's memory, which is not timeless eternity, but everlasting life, in which everything is *alive*. Schmemann develops similar themes to Johnson when he considers the place of memory in the celebration of the Eu-

69. Schmemann, *The Eucharist*, 125.
70. Schmemann, *The Eucharist*, 125–126.
71. Schmemann, *The Eucharist*, 126.
72. Schmemann, *The Eucharist*, 129.

charist. In bringing the gifts to the altar we also 'include the ones being remembered in the life-creating memory of Christ: God's memory of man, man's [*sic*] memory of God, that divine memory which is eternal life'.[73] This memory is active and participatory, because 'we *create the memory of each other*, we identify each other as living in Christ and being united with each other in him'.[74] In other words, by our love for one another and our commemoration of one another, we bring one another to Christ in prayer; we hold each other up to God. All of this is accomplished in the celebration of the Eucharist, in which we offer a sacrifice, through the sacrifice of Christ. There is no distinction made between the living and the dead, or, I would add, between the human and the non-human. This is because, in remembering them and bringing them into the 'life creating memory of God, the borders between the living and the dead are erased, for all are recognised and *manifested* as living in God'[75] [my italics]. The sacrifice of Christ on the cross, made once and for all, is made manifest in our lives through the offering up of our memory of Christ and of one another.

The final eschatological transformation of all things is happening when we bring creation into the memory of God. Our participation is significant. The dead are manifested as living in God through our remembrance of them. But the living are also caught up in this process, including *all* that lives. It's important to stress at this point that any definition of living, which does not include Earth itself, in its entirety, and indeed the whole created order in its entirety, is limited and, from an eschatological perspective, defective. *All* that exists needs to be remembered by us and brought to God, that all may live in God.

8.8.4 The sacrament of anaphora

Schmemann reiterates his conviction that the liturgy must not be reduced to 'moments', in the sense that theologians attempt to identify 'the moment' when, for example, the bread and the wine become the body and blood of Christ. Having said that, the language of the eucharistic canon focuses the worshippers' hearts and minds on the chief meaning of the Eucharist. However, Schmemann continually repeats that the divine liturgy must be seen as a whole, with the unified purpose of continually lifting up the church, and with it the whole of creation, 'to the unfading light and

73. Schmemann, *The Eucharist*, 130.
74. Schmemann, *The Eucharist*, 130.
75. Schmemann, *The Eucharist*, 130–131.

joy of the kingdom of God'.[76] Liturgy is therefore understood as precisely the literal meaning of its root word (*ergon* in Greek), that is, an action.

Within that context, the three verses and responses at the start of the eucharistic canon help shape the rest of the prayer. Firstly, the solemn blessing, mostly commonly, 'The Lord be with you'. The Christian faith, and the Eucharist in particular, begins with an encounter with Christ and, through Christ, with God the Holy Trinity. The Eucharist is the source, *par excellence*, of the church's encounter with the Trinity.

Secondly, the invitation to 'lift up our hearts', as Schmemann points out, is a phrase exclusive to the divine liturgy. This is an invitation to ascend to the 'real homeland of our heart's desire, to which we returned after an agonising exile, for which we have always groaned with homesickness, and through the memory of which all creation lives'.[77] The Eucharist elevates us to the heavenly kingdom. Heaven[78] transfigures the earth in the celebration of the Eucharist and the earth understands heaven as the ultimate truth about itself.

Thirdly, the phrase 'Let us give thanks unto the Lord' is an invitation to join Christ in offering thanksgiving to God. I shall now examine the significance of the profound liturgical act of thanksgiving.

8.8.5 The sacrament of thanksgiving

Schmemann's work depends on the understanding that the act of thanksgiving has ontological significance. It is not an empty action, in the sense of being purely an emotional response. Rather, consistent with Schmemann's theology of symbolism, the act of thanksgiving is the means by which we *accomplish* the remembrance of Christ. Every time we give thanks for and bless the world, we 'transform it into what it was created for and granted to us by God'.[79] In accomplishing the remembrance of Christ, we see the world as the new creation, 'in which everything created by God is called to become our partaking of the divine love, of the divine life'.[80] Each time we give thanks, each time we raise the world up to God, through the remem-

76. Schmemann, *The Eucharist*, 165.
77. Schmemann, *The Eucharist*, 168.
78. It's important to note that Schmemann is using the category of 'heaven' here as meaning the relative transcendence of the earth; it is part of creation. Heaven and earth will then be transformed into the new heaven and the new earth, the kingdom of God, in which 'God will be all in all'. Schmemann, *The Eucharist*, 169.
79. Schmemann, *The Eucharist*, 177.
80. Schmemann, *The Eucharist*, 177.

brance of Christ, the salvation of the world is made complete; we and all of creation are raised up to paradise in the act of thanksgiving.

Crucial to Schmemann's theology, and reflecting his Orthodox tradition, is the belief that in the eucharistic act of thanksgiving the church partakes of the divine nature. The Eucharist is above all a sacrament of the knowledge of God the Father. Moreover, every act of thanksgiving in this context exposes the truth of sin and evil, which is forgetfulness of God and of humanity and creation's true nature: that we were created for communion with God. When we fall away from that calling, to be in communion with God, we cease to remember our true nature. But the very act of thanksgiving restores this to us. Every act of thanksgiving lifts us up to the throne of heaven, to the Father. We know God and have communion with him. Embracing all of creation, a point which Schmemann continually repeats to his readers, the church is lifted up to heaven and thus becomes heaven on earth.

In this context, to draw out the specifically eschatological meaning of Schmemann's work, every act of thanksgiving is an eschatological act. Every time we give thanks to God, that is, raise the world up to God, we participate in the inauguration of the eschaton. Everything, not just the church herself, is lifted up to God and thus heaven is made manifest on earth.

8.8.6 The sacrament of remembrance

The Eucharist elevates us to the throne of heaven, because it is a sacrament of remembrance. We do not somehow elevate ourselves, as though salvation were accomplished through the power of human beings, rather we are elevated through the remembrance of Christ. We experience the very reality of the kingdom of God, because we remember, in the sense of comprehend it *as real*, the kingdom which Christ made manifest at the last supper. Schmemann understands the Eucharist as knowledge of the last supper, which was 'the ultimate manifestation of the kingdom of God'.[81] Importantly for atonement theology, with its focus on the sacrifice of Christ on the cross, Schmemann does not separate the last supper from the events of Good Friday or Easter Day. The same divine love, which is 'the very essence and glory of the kingdom of God',[82] which was made

81. Schmemann, *The Eucharist*, 201.
82. Schmemann, *The Eucharist*, 205.

manifest and granted at the last supper, also transforms the cross into victory.

The whole liturgy is a remembrance of Christ, specifically of Christ's sacrifice on the cross, which in Christ becomes *ours*.[83] We too participate in the suffering of Christ. Yet, through Christ, we are given a part in the kingdom of God. Through the remembrance of this sacrifice, the kingdom of God, in the power of the Holy Spirit, is made manifest and fulfilled.[84]

8.8.7 The sacrament of the Holy Spirit

So we reach the climax of the Eucharist in the *epiklesis*, the invocation of the Holy Spirit, which completes the eucharistic celebration. Schmemann is once again quick to point out that the failing of both Eastern and Western eucharistic theology has been to attempt to identify a consecratory moment or formula. In the East this formula became the epiklesis, which was understood to be the cause of the change of the holy gifts into the body and blood of Christ. But the entire liturgy needs to be understood as a sacrament and that, through its many parts, the Eucharist is accomplished.

The invocation of the Holy Spirit transfigures the old earthly time into the new time of God's new creation. The world is not replaced, but rather, through the Holy Spirit, we see and receive the world 'as God created it: heaven and earth, full of the glory of God'.[85] Schmemann's use of language is particular. Although the terms are by no means completely interchangeable, the words 'remembrance', 'manifestation' and 'epiphany' all describe the essence of the liturgy. These terms all express his conviction that the salvation of the world, accomplished by Christ, is made present, made real, in the celebration of the Eucharist. 'And thus each liturgy is a *gathering*, a restoration and an "identification" of the fullness of our remembrance.'[86] Whilst the liturgy is served on earth, it is accomplished in heaven. We are raised up in the Holy Spirit and in the Holy Spirit the old time is transfigured into the new time. The entire liturgy then becomes a transformation, by which Schmemann means that each part of the liturgy, each rite, 'is transformed by the Holy Spirit into *that which it is*, a "real symbol" of what it manifests'.[87] The Holy Spirit then unites all those things

83. Schmemann, *The Eucharist*, 211.
84. Schmemann, *The Eucharist*, 211.
85. Schmemann, *The Eucharist*, 219.
86. Schmemann, *The Eucharist*, 221.
87. Schmemann, *The Eucharist*, 223.

which have come to pass, that is, the 'entire mystery of salvation accomplished by Christ, the mystery of Christ's love, which embraces the whole world and has been granted to us'.[88]

In the celebration of the Eucharist the work of Christ is not repeated. Schmemann is at pains to point this out: the celebration of the Eucharist is not a repetition of the saving work of Christ. This is not the meaning of remembrance. Rather the 'rites of the liturgy are a manifestation, one after the other, of the *realities* of which the saving work of Christ is comprised'.[89] The epiklesis is the manifestation of these realities and the church's acceptance of them.

8.9 The cosmic liturgy

Using Schmemann's sacramental theology and comments about symbolism as a model, the Eucharist itself ought not to be understood as being isolated from the rest of creation. If, within the context of the Eucharist, the epiklesis is the climax of the celebration, but not somehow separate from the rest of the liturgy, then the same is true of the Eucharist. It is the climax of the liturgy of God's creation, the cosmic liturgy, but not separate from it.

Thus, lest this eucharistic theology be viewed as being overly anthropological, it is important to explore the deep connections between the church, the priest and the rest of creation in Eastern Orthodox theology. The Eucharist is never celebrated in isolation from the rest of God's creation. Schmemann quite helpfully writes that, 'The Church is not a religious cult but a *liturgy*, embracing the entire creation of God'.[90] The Eucharist is understood as a 'cosmic sacrament'[91] and the world 'as the "matter", the material of one all-embracing eucharist'[92] in which human beings are the priests. Human beings receive the world from God and then offer it to God, in an act of blessing God. They do this on behalf of the world, connected with it. Therefore a sacrament, particularly the sacrament of the Eucharist, is both cosmic and eschatological, because it refers both to the world as God created it and 'to its fulfilment in the kingdom of

88. Schmemann, *The Eucharist*, 224.
89. Schmemann, *The Eucharist*, 225.
90. Schmemann, *The Eucharist*, 242.
91. Schmemann, *The World as Sacrament*, 16.
92. Schmemann, *The World as Sacrament*, 16.

God'.[93] The question of whether Schmemann's work nevertheless remains too anthropological, because of the emphasis on the world as the 'matter' of the cosmic sacrament, has been explored by Elizabeth Theokritoff. She considers whether there are 'less passive images for the role of non-human participants in the cosmic liturgy'.[94] It is a significant question, which I will address briefly in the conclusion of the book.

8.10 Conclusion

As living symbols of Christ in the community around us, we play our part in the transformation of the cosmos. We are not removed from this world, with all its struggles for justice (environmental and social). Rather, filled with the hope that God will bring his transformation of the cosmos to completion, we participate in the process of this occurring, particularly through the celebration of the sacraments. What this book as a whole has attempted to show is that concern for justice, ecology and personal fulfilment in this world does not mean that eschatology must be rejected as a theological hope. In this chapter, through an exploration of Alexander Schmemann's theology, I hope I have shown an alternative to Keller and Ruether's approach (and indeed also Körtner and Moltmann's). God uses the processes of this world to bring about the consummation of all things. We too are invited to participate in that. Using the focus on memory in the theology of Johnson, and then the sacramental theology of Schmemann, I have suggested in this section that, in the use of memory, and the celebration of the sacraments generally and in the Eucharist in particular, the church as the community of believers is caught up into the mystery of Christ's life, passion, death and resurrection. Through the ritual celebration, the remembrance of Christ's actions, we proclaim to God and the world the mystery of Christ's sacrifice and so are enabled to participate in the transformation of all things in Christ.

This understanding of eucharistic theology provides a model for our participation in the transformation of creation and, I am suggesting, the remembrance of creation, both by us as human beings and by God. Moreover, our praise of God, on behalf of, and in connection with, all of creation, calls and challenges us to act ecologically. The vision of a transformed creation calls us to make a commitment to participating in that

93. Schmemann, *The Eucharist: Sacrament of the Kingdom*, 34.
94. Elizabeth Theokritoff, 'Creation and Priesthood in Modern Orthodox Thinking', in *Ecotheology* 10.3 (December 2005): 347.

transformation. This is our ethical response to the joy of praising God with all of creation. An ethical response must go hand-in-hand with an eschatological vision, as Carol Dempsey reminds us:

> The prophetic vision of new creation is a false and empty hope unless it empowers us to embrace new ways of living on this planet that anticipate the future. Indeed this will require grace, but it will not be cheap. Hope will arise from the eschatological horizon of faith and the day-to-day commitment to do justice and righteousness to all creation.[95]

The praise of God, and the offering up of creation to God, must therefore serve as an inspiration to action.

An examination of Schmemann's work on the sacraments of the church illustrates how the understanding of human beings as priests of creation might function. All of creation longs to be 'restored'[96] to the purpose for which it was created. That is, it longs to be restored to its role of praising God, and in doing so, revealing God. This is the true role of all of God's creation. When all of creation is engaged in its true role of revealing God, it is participating in the transformation of all things. For example, the blessing of the water in the sacrament of Baptism restores water to its true function.

Schmemann's work on the Eucharist is the prime example of how we might understand the process of giving creation back to God. As a sacrament of the kingdom, the Eucharist is not merely a memorial meal but rather communicates the reality of the kingdom and allows us to participate in its manifestation. The various parts of the Orthodox liturgy, which are reflected in other eucharistic traditions, reiterate the meaning of the Eucharist as the sacrament of the kingdom. In this chapter I concentrated on those parts of the Eucharist most significant for eschatology.

All of Schmemann's work points us to an ecologically acceptable model of eschatology, which is to view the processes of creation as a cosmic liturgy. Through the ritual celebration, the remembrance of Christ's actions, we proclaim to God and the world the mystery of Christ's sacrifice and so are enabled to participate in the transformation of the cosmos. This cosmic liturgy of all of creation restores the world to what it was once

---

95. Dempsey, 'Hope Amidst Crisis: A Prophetic Vision of Cosmic Redemption', 282.
96. See my comments on the question of temporality with reference to the concept of the 'fall' and the use of the word 'restore' above, at footnote 97.

was, and is meant to be, a sacrament of God's presence. The Eucharist is the climax of what is happening in the rest of creation. It is also intimately connected with what is happening in the rest of creation, and therefore shapes our ecological action within that creation.

Schmemann's liturgical theology gives us a framework in which human beings can be seen to be active and participating in the process of what Conradie terms 'material inscription'. Both Johnson and Schmemann help us by showing us that human beings can indeed participate in God's transformation of all things in Christ.

Conclusion

The basis for our belief in bodily resurrection, suggests John Haught, 'is the compassion of God, God's inability to forget what has happened to us and the world. The power of God to raise us from the dead is of a piece with God's loving embrace of the entire cosmic story which, in perishing, passes into the everlasting divine feeling.'[1] God compassionately saves all events everlastingly in the divine experience. So, suggests Haught, our personal destiny is never separated from the entire cosmic story. If we are able to think of our bodily resurrection as being completely inseparable from the ultimate purpose of the entire universe, then we might be more predisposed to act in an environmentally responsible way, because we will be related to the natural world forever, with all its diversity and complexity.

In this context, to return to Elizabeth Johnson's work, if we break the connection with the memory of the dead and lose hope for them, it can in fact 'have deleterious effects on our ecological responsibility',[2] because we assume that there is an unbridgeable gap between the dead and ourselves. The severing of the connection between the generations undermines the basis for ecological ethics, because we are stranded, as Haught suggests, 'in a meaningless brief life span'.[3] But a shift in the understanding of the communion of saints addresses this dilemma and widens our vision to include not just those human beings who have gone before, but the whole community of the natural world as blessed by God and drawn into God's eternal presence.

1. Haught, *The Promise of Nature*, 142.
2. Johnson, *Friends of God and Prophets*, 242.
3. Haught, *The Promise of Nature*, 129.

1. The participation of the rest of creation in the transformation

One of the limitations of using the Eucharist as a model of how we participate in the transformation of all things, is that it is potentially an anthropocentric model. However, as long as it is understood to be *one* example of how the processes of creation are valued, then its focus on *human* participation in the transformation of all things is not critical. Alternatively we might consider the Eucharist the crowning and the fulfilment of what is happening in the rest of creation. Within the cosmic liturgy, the Eucharist occupies pride of place as the unique gift of Jesus to his disciples, as a means of offering all of creation back to God.

If, though, the Eucharist is the climax of what is happening in the rest of creation, the question then becomes how all of creation participates more actively in the process of transformation, or, in Conradie's terms, the 'material inscription' of everything. Certainly the connection between human beings and the rest of creation cannot be overstated in our current climate of ecological abuse and wealth creation partly based on the exploitation of natural resources. Lowes suggests that there is 'an ineradicable reciprocity'[4] between human persons and created non-personal beings. Indeed human persons preserve their own personal substance, that is, their very personhood, by relating to other beings, both personal and non-personal. Human beings depend, at least partially, on their relationship with non-personal beings to define their own personhood. The reciprocity of this relationship is heightened in the eschaton. He suggests, in the eschaton the 'transfiguration of the non-personal universe is as necessary to human persons transforming within the power of the resurrection of Christ within and through death, as they to it.'[5]

Lowes is saying here that human beings will only exist through and beyond death within the framework of the 'new heaven and the new earth'. But there is another consideration as well. If human beings do indeed have a participatory role in the transformation, as I'm suggesting they do, then this old creation must also be allowed to play its part in its own transformation. In other words, the processes of creation, and the non-personal beings within creation, have a role to play in the eschatological transformation of all things. This point is a development of what Lowes is suggesting. His critical point, which is still somewhat anthropological in focus, is that human persons contribute both to their own transformation and

4. Lowes, 'Up Close and Personal: In the End, Matter Matters', 141.
5. Lowes, 'Up Close and Personal: In the End, Matter Matters', 141.

that of the wider material milieu (within the larger transfiguring work of the Trinity) 'to the degree that they have managed to love them [the non-personal created beings] unconditionally and serve them selflessly'.[6] Of course a significant factor in loving non-personal created beings unconditionally is undertaking not to exploit or destroy them. Moreover, the realisation that they have a role to play serves as a significant reason to allow them to do just that. In other words, rather than being destructive towards the rest of creation, we allow it to play its role in the transformation of the world. An ecological ethic emerges from the realisation that the rest of creation participates in God's work.

If the relationship between human and non-human beings is indeed reciprocal, exactly what part do the non-human beings play in transforming the world? A potential model may be found in psalm 148. This song of praise, linked to the song of the three young men in the fiery furnace, is a song which all of creation is called to sing. Not only are living creatures, including human beings, called to praise God, but every created thing. From the sun, moon and stars, to the sea-monsters and all deeps, to the fire, hail, mist and snow, to the mountains and trees, all created things are called to praise the Lord. It is this litany of praise which unites all created things in praising God's glory, as Weiser puts it: 'The glorification of the Creator and Preserver of the world fulfils the ultimate depth of meaning which unites the inanimate created things and the living creatures in a mutual relationship; to praise the sole majesty of God is the final goal which unites the whole universe in a communion of God's service'.[7] Thus our relationship with all created things is strengthened and expressed most perfectly when we join with all created things in God's praise. For human beings this praise finds a particularly powerful and beautiful expression in the celebration of the Eucharist. How then do other created things sing God's praises?

Other created things sing God's praises precisely when they embody everything God means them to be. When they are the fulfilment of all that God has created them to be, they are singing God's praises. When the wedge tail eagles, some of the largest birds of prey in the world, soar in the clear Australian sky thousands of feet up in the air, their keen eyes searching for movement on the ground, they are the embodiment of everything God has created them to be. When the autumn rains fall, bringing life and health to every living thing, and the sunburnt Australian countryside be-

6. Lowes, 'Up Close and Personal: In the End, Matter Matters', 143.
7. Artur Weiser, *The Psalms: A Commentary* (London: SCM, 1962), 837.

gins to turn lush and green, then we see the natural world itself being the embodiment of everything God has created it to be. All of this and more is part of the cosmic liturgy. If this liturgy sounds discordant or rough to our ears at times, it is because the music of God's cosmic liturgy is different from what we might expect.[8]

Human beings, because of sin, always fall short of what God has created them to be. The importance of soteriology cannot be underestimated in eschatology, as Conradie reminds us. This is true of the natural world too. Using the terminology of Romans, 'we know that the whole creation has been groaning in labour pains until now' (Rom 8:22). And in this very groaning, creation itself participates actively in the process of 'material inscription', because it sings God's praises. It can be argued that human beings are the ones who have upset the sacred balance.[9] But, at our best, when we participate constructively in the cosmic liturgy, human beings *remember* all of creation, both in the celebration of the Eucharist and in other acts of remembrance, and thus bring creation and themselves to God. Creation longs for human beings to do just that. As profoundly spiritual creatures we commend all of creation to God and bring it to God. But the nature of the reciprocity means that creation itself also plays a role.

This is the participation of created things in the transformation of the world. Crucially, the transformation must involve, at a fundamental level, the work of the triune God, who is both immanent and transcendent. Without God's work there can be no transformation. But, equally, the singing of God's praises, whether in the Eucharist or in the wider created order, and the *memory* of all of this, allows human beings and all of creation to participate in the final consummation of all things.

Elizabeth Theokritoff has taken Alexander Schmemann's eucharistic theology and moved further in this understanding of the cosmic Eucharist. She points out that this vivid sense of the creation as a whole responding to its Lord finds expression in Orthodox worship in the prominent use of psalms such as 148 and the Song of the Three Children (the use of both of which I explored above) as well as psalm 104 and many other scriptural texts. This is the basis 'for the Orthodox notion of cosmic Eucharist and of the human being's role therein: *the praise offered by all creation is not an end in itself*, but a mode of being *intended to culminate in the transfigura-*

8. I am indebted to my friend and colleague the Rev'd Dr Ian Mackay for this point.
9. For a discussion of this balance see David Suzuki with Amanda McConnell, *The Scared Balance: Rediscovering our Place in Nature* (St Leonards NSW: Allen & Unwin, 1997).

tion of all things in Christ.'[10] Thus the role of other creatures is more than a passive role. They are not only 'the substance of the offering, but also concelebrants in the offering of glory to God'.[11]

Theokritoff reminds us that the creativity of human beings can serve transfiguration, rather than disfigurement. To use Christopher Southgate's language, human beings can be 'co-redeemers with Christ of the whole evolutionary process'.[12] This role is not restricted to human beings, but includes other creatures, which is precisely Theokritoff's point.[13] I would add that we play a far greater role than we realise when we simply allow the rest of creation space to praise God. Just as God allowed space for creation to come into being, and to develop and evolve, so too we need to allow the non-human part of creation the space it needs to sing God's praises. The concept of the preservation of the world takes on new meaning, when we realise that the rest of creation needs the space to play its participatory role in the transformation of all things. This becomes a powerful reason to be ecologically responsible.

The singing of God's praises is the eschatological goal, as Moltmann reminds us: 'The glorification of God is the ultimate purpose of creation'.[14] If we understand God's glorification in trinitarian terms, 'the mutual relationships of the Trinity are so wide open that in them the whole world can find a wide space, and redemption, and its own glorification'.[15] Extending Moltmann's thoughts here in a direction he admittedly does not take them, the very glorification of God by creation means that its own glorification becomes possible in the open Trinity of love. The very singing of God's glory then, in turn, becomes our way of participating in the divine glory. As long as the transcendent nature of God's salvific action is also emphasised, this model of singing God's praises then becomes the participatory or subjective side of the model we have been searching for. If 'material inscription' is the objective, transcendent side of our model of ecological eschatology, then liturgy is the subjective, immanent side. With all of creation we sing God's praises, we remember before God those things past, and we look with hope to the future life in the triune God.

10. Theokritoff, 'Creation and Priesthood in Modern Orthodox Thinking', 357.
11. Theokritoff, 'Creation and Priesthood in Modern Orthodox Thinking', 360.
12. Christopher Southgate, 'God and Evolutionary Evil: Theodicy in the Light of Darwinism', in *Zygon* 37 (June 2002): 818.
13. See also Elizabeth Theokritoff, 'Creation and Salvation in Orthodox Worship', in *Ecotheology* 10 (2001): 97–108.
14. Moltmann, *The Coming of God*, 323.
15. Moltmann, *The Coming of God*, 335.

My conclusion is that the rest of creation is longing to become everything that God wants it to be. Creation itself participates actively in the process of 'material inscription', because it sings God's praises. Human beings have upset the balance. But, at our best, when we participate constructively in the cosmic liturgy, human beings *remember* all of creation, both in the celebration of the Eucharist and in other acts of remembrance, and thus bring creation and themselves to God. Ultimately the triune God is the one who transforms the creation into the purely sacramental world it was intended to be. To use one of Moltmann's (and Paul's) favourite phrases, 'God will be all in all'. But our participation in that process of transformation, or 'material inscription', is crucial, and is a compelling reason to behave environmentally responsibly.

Biblical References

Index